Robert Vaughn

Fine-Tuning Your Writing

Advanced writing techniques for scientists, engineers, physicians, and other professionals

William R. Luellen
Fine-Tuning Your Writing, LLC.

2001

WISE OWL
Publishing Company

First edition, 2001

05 04 03 02 12 11 10 9 8 7 6 5 4 3 2

Wise Owl Publishing Company is a division of
Fine-Tuning Your Writing, LLC.
929 S. Midvale Boulevard
Madison, WI 53711

Library of Congress Control Number: 2001090393

ISBN: 0-9709415-0-1

SAN: 253-7680

Printed in the United States of America

Contents

Acknowledgments

No book happens in isolation. And although it sometimes seems like it, no writer writes in isolation. There are a great many people who could be cited here, and more than a few that should. I will try to mention all the *shoulds*, and hope the *coulds* will realize that while their name is not mentioned, it will always remain in my memory.

For instance I mention in the book that it all began with an unnamed author who had incorrectly alphabetized a reference list. That's a little like saying that World War II began when Lt. Cmdr. Kakuchi Takashshi dropped a bomb on Hanger 6 at the south tip of Ford Island in Pearl Harbor, Hawaii. Obviously a great deal had to happen before that instant to get his Aichi D3A1 Type 99 "Val" Dive Bomber (not to mention several hundred more planes and dozens of ships) into position to drop that bomb. And, of course, a great deal of fighting had been going on in Europe before the leaders of Japan decided to attack America.

Why the analogy to war? Because in a sense I've been fighting bad writing all my life. The earliest incident I can specifically recall was reading an article in the student newspaper at my high school and thinking, "I can write better than that." I never wrote for that high school newspaper, but I had written for the student newspaper in junior high school and wrote and edited for several publications while in college.

I definitely should mention Richard Hebert, an early teacher who became a friend and Evelyn Heden, an early friend who became a teacher as two people who had the strongest influence on my early life helping me see who I was and what I could become.

Then there was Carl Hamilton, a former newspaper editor who headed the journalism department at Iowa State University when I declared my major there, and John Stowell, the toughest editor I ever worked for. I learned more from him in 3 months on my first job than I had learned the previous 3 years at Iowa State. I also thank Matt Stelly for taking a chance and hiring a guy from a newspaper to take over a magazine, Harry Hamilton who taught me much, and Domenic Fuccillo, with whom I had many conversations about writing quality. Richard Dinauer was another early colleague who along with Susan Ernst, Jon Bartels, Jay Poster, and Dave Kral served as the peer-review team on the original columns. The same Dick Dinauer, more than 30 years after I first met him, proofread the final copy of this book.

Jerry Hatfield and Gary Peterson gave me a great deal of assistance through the years and even acted as co-authors of a couple of the columns in this book. Dwayne Buxton and Steve Thien answered a lot of questions for me through the years, as did professional editors like Pamm Kasper, Nich Rhodehamel, Barb Seybold, and many others.

I cannot count the number of scientists who encouraged me to write this book. If it were not for them, though, quite likely there would be no book. Then more scientists stepped forward with their money when I made the announcement I would write the book, but would need several hundred prepaid orders to finance its printing. One scientist who deserves special mention is Kazuhiko Kimura, who took it upon himself to become my sales agent in Japan and sent me a prepaid order for more than 150 copies. With that order I achieved the critical mass that allowed me to move forward with its printing and publication.

Special thanks go to my wife of 38 years, Nancy, for putting up with the long hours I spent with my computer creating this book instead of with her. Thanks, dear, now that it's done, let's have a nice dinner together before I begin working on my next book.

And thank you, God, for creating all the above and bringing it all together at the proper time and place.

—William R. Luellen

Introduction

No doubt you have seen colleagues or students handle themselves so well in a field, laboratory, classroom, or giving a speech that you remark, "That person really knows what he (she) is doing." Editors notice that about authors as well.

The way you write and your attitude toward other people's suggestions and criticism will create a halo (either a good or a bad halo) for yourself in the minds of editors. Savvy authors will try to create a good halo by writing and acting in a way that indicates they are pros who know what they are doing.

Such authors are rewarded for their efforts; editors eagerly jump at the chance to work on their papers, often setting aside other work, because they know the paper will be well written, interesting to read, and informative. They also know the author will be cooperative.

The by-lines of authors with bad halos evoke groans: Editors have been known to put off working on papers by these authors for several days until they are able to clear the long uninterrupted time period they know the paper will require.

The previous paragraphs were the opening of an article published in February 1994. That article, which appears in its entirety in Chapter 3 of this book, was one of a series of columns under the general title of "Fine-tuning your writing" I wrote for 8 years and had published in the member newsletter (then called *Agronomy News*) of three scientific societies, the American Society of Agronomy, the Crop Science Society of America, and the Soil Science Society of America, headquarter in Madison, WI. (In this book I either abbreviate those organizations as ASA–CSSA–SSSA or simply call them the Societies.) Many of those original columns were also reprinted in the newsletters of other organizations and societies across the nation and around the world.

Near the end of each volume year, all the original columns were compiled into an annual booklet. The columns in these reprint books were in the exact form they were published, including the changes by some editors that I was not entirely satisfied with. The book you have in your hands now, in part, is a compiling of those original columns (except two or three that have gone out of date) in a form closer to the way they were originally intended. But this book is not merely a shoveling together of a bunch of old columns. I tried to do five things as I created this book.

1. Every column was reread and all that needed it were updated.
2. All were grouped by subject matter, not by date published.
3. A (usually) brief comment from me proceeds, follows, or occasionally appears within each column.
4. New information was added wherever needed.
5. Along with those updated original columns, this book also contains material from my *Fine-Tuning Your Writing* newsletter, which continues the work began in the original columns, and from other sources that have not appeared in either the original columns or my newsletter.

In short, this book contains the best of the best information from the original columns, the new newsletter, and other material that has not appeared in either. Each chapter opens with a new introduction from me. Then, depending on the nature of the chapter and the information it contains, that introduction is followed by one or more entire columns from the original series, portions of one or more columns from the original series, and sometimes also information from other sources.

In short, I believe this is the most up-to-date book on the subject of writing, especially the writing of the American version of scientific and technical English.

Who needs this book? You do, if you are (or want to be) one of those writers referred to at the beginning of this introduction—the sort of person who writes material that editors look forward to reading.

I have found in my 40-odd years in this business that the two groups of people most likely to ask me for advice are (i) writers who already are excellent at what they do, (ii) and writers who are good at their writing and moving in the direction where they soon will be classed among the excellent. Interestingly, poor writers generally disdain my help with a curt, "I already know how to write." Technically, what these people say is true; they *do* know how to write. But they don't know how to write very well, and because of their attitude, they are never going to get any better than they are now.

For those who have never heard of *Fine-Tuning Your Writing*, I'll take a couple paragraphs to explain its origin. Sometime in early 1992, after editing several manuscripts in a row whose authors had made similar errors in their Reference sections, I screamed to nobody in particular, "Doesn't anybody know how to alphabetize a reference list!?"

Immediately I took a break from this editing and wrote a couple-hundred word essay titled, "How to alphabetize a reference list."

Just before I handed that essay to the editor of what was then called *Agronomy News*, the membership newsletter for the above-mentioned three scientific societies, I placed the words, "Fine-tuning your writing" above that title. To this day I am not completely certain why I added those words. In part, though, it was because I envisioned maybe someday writing another essay or two on the same general subject.

That "some day" became the next month, and the next month, and the next month until I had written 93 consecutive articles under that overall title.

By that time, 8 years had passed and I was ready to retire. My former employer did not want to continue to publish the column, so I obtained the rights to it and took it with me. That column quickly evolved into a monthly newsletter that is one of the services offered by my writing-editing-consulting business called Fine-Tuning Your Writing, LLC. (The LLC stands for limited liability company—something like a corporation.)

If you have not heard of Fine-Tuning Your Writing, you likely also have not heard of me, so now I'll take care of that. I'm the retired managing editor of three scientific journals. I held that position for about 14 years. For 16 years before that, the same publisher employed me as the editor and usually the only writer of an extension–outreach magazine, called *Crops and Soils Magazine*, which had an international circulation.

Before my 30-year career with that employer I worked as a reporter, copy editor, for a daily newspaper in Wisconsin, and before that, I had a similar position with a daily newspaper in Iowa.

To date I have written close to 4 million words for publication. On top of that, in my roles as editor, managing editor, and now publishing consultant, I have helped many people learn how to write and how to edit. In my spare time I write nonfiction feature stories, which I have published, and fictional short stories and novels. I've even written some music for special events.

In short, the information in this book is based on the knowledge and experience I've gained from more than 40 years of writing and editing, helping others write and edit, discussions with colleagues, and reading.

Is this book the last word in writing? I hope not. If it is, that means writing will never be better than it is now. I hope not only my own writing, but also that of others will continue to improve for as long as people continue to write.

Some who like to think of themselves as my detractors have gone so far as to say that the material I write in my columns and newsletter is my opinion only, and not necessarily related to the facts as they should be presented. I plead guilty to the first part of that criticism. Of course it is my opinion. But that opinion is based on 40 years of working in the business of putting my own words together *and* helping others put their words together so they sound and read like their words and not mine.

As for the second part of the criticism, perhaps a tiny bit of it is true. I have a standard answer that applies not only to the material I write, but to all material I read. That answer is not my own creation, but the words of a rabbi named Paul, who wrote them close to 2,000 years ago: "Test everything, and hold onto what is good."

I feel confident that any good writer who tests the words in this book will find that many, if not most of them, are good.

More than that, if you write, and particularly if you are careful about the way you write, I think you will find the information in these pages will be valuable to you, regardless of your current level of experience or talent.

William R. Luellen
Madison, Wisconsin
6 April 2001

Chapter 1

Consider your reader

Like all books on writing, this one begins with the basics. But unlike most of those books (and authors) I do not believe the most basic aspect of writing is a collection of arcane grammar and spelling rules. The true basic, I believe, it is to have a firm notion in mind of who you are going to write for and what you want to tell those persons.

Of course grammar is important, and I'll get to it in Chapter 6. Of course spelling is important, but if you don't already know how to spell this book isn't going to help you, so it contains nothing specifically on it. My firm belief is that a perfectly punctuated, grammatically correct, perfectly spelled piece of writing is worthless if it does not tell the target audience what that target audience needs to know in the sort of language that audience can understand.

Thus, the first five chapters of this book deal with what I have found to be the true fundamentals of writing. And the most fundamental, most basic rule of all for writing is expressed in the words that make up the title of this chapter. When you are writing something to be read by native-born-American scientists with Ph.D. degrees, do not use the same language you would when writing for Ph.D.-degree scientists for whom English is a second or third language. When you are writing for undergraduate students, do not use the same language you would use for native-born-American scientists with Ph.D. degrees.

The same rule applies when you are writing for nonscientists, whether they be children, homeowners, senior citizens, or anyone.

What do I mean by the phrase "same language?" Maybe it would be easier to describe what I *do not* mean. Of course this book deals with documents written in English, or whatever language might be used by the writer and reader. I am referring to a host of variations within that language that make it "the language" used by a particular group of people. Those variations include such things as length and complexity of words, length and complexity of sentences, and the inclusion or lack of inclusion of words understood by one particular group and nobody else.

Beyond those basics are a host of considerations, ranging from what you, as an author, want to convey to what the reader wants and needs to know. You may find this difficult to believe, but both of these—the author's needs and the reader's needs—are closely intertwined. Perhaps a good way to explain what I mean would be to reproduce three columns that were first published in March, April, and May of 1995.

The intended audience for these three essays was scientists or students studying to become scientists. I'll present more universal comments and concepts after these three papers.

What does your reader know and need? (part 1)

Authors and editors often call and ask me how to handle some particular point in a paper they are working on. This is certainly fine with me; part of my job, in fact. But they could find the answer to many of those questions without me if they would ask themselves one question, "What is the best way to present this information to the reader?"

Before that question can be answered, though, one must consider another question; "Who will read this paper?" Readers of a scientific journal or book can generally be grouped into four categories: (i) experienced scientists fluent in English, (ii) experienced scientists not so fluent in English, (iii) not-so-experienced scientists who may or may not be fluent in English, and (iv) late undergraduate or early graduate students, who may or may not be fluent in English, and who are just learning about your branch of science.

How can a person write for such a diverse clientele? You can't, but you don't have to. An author of scientific English has the goal of writing clearly and completely enough so that a *colleague or peer reviewer* can understand and follow what is written. In other words, write for Group (ii) or (iii). If you do that, you will automatically reach Group (i) as well. Papers written this way will not exclude Group (iv) either; undergraduate students will not be working on their own; a major professor or advanced graduate student will help them.

Someone once came up with the three Cs of writing; clear, concise, and correct. These are good as far as they go, but I believe there is a fourth C that is just as important; complete. What absolutely has to be in a paper and what may be left out? The answer to this question can also be found when you think about the reader. If you keep the reader in mind and follow all four Cs, your readers will be able to say, "Now *here's* a paper I can understand!"

Examples

For instance, how do you cite a long-out-of-print publication? Think of the reader, who would ask, "Where can I get that publication?" Your answer might be that a substantial portion of it was reprinted elsewhere or that a newer publication has superseded it. If so, tell your readers about it in your reference section. If the paper is old, but still available in one form or another, let the reader know where. If the cited paper is both not available and not copyrighted, consider making your own copies and letting the readers know you will supply one to any who request it.

Another set of questions I'm often asked concerns publishing the name and address of the manufacturers of products mentioned in a paper: Is this always essential? Again, think of the reader. If a paper deals with a soil sampler that you mounted on a Ford pickup truck, you would probably identify the soil sampler, but would generally not need to say that the truck is manufactured by the Ford Motor Company in Dearborn, MI. You probably would not have to even say it was a Ford pickup. For instance, if the important thing is the size and weight of the truck, not the brand, you would be better to define the need instead of using a brand name ("this sampler is heavy and requires a great deal of power; it must be mounted in a truck with 4-wheel drive and at least 250 horsepower").

The same is true for laboratory equipment, most of which is readily available in any number of catalogues and from numerous supply houses. The brand name needs to be used if it is the only one that will work in a particular experiment. If you use an obscure brand not in widespread distribution or a piece of software you developed yourself, tell the readers where they can buy it.

I once dealt with a problem that combined the two problems just mentioned. A writer called and asked me how to describe a product that was essential to the work, but no longer commercially available. The author also mentioned that there was a similar product on the market that could be substituted for the product he used. That was all I needed to answer his question. I suggested he use a sentence along the lines of, "The Acme no. 2000 Widget that we used is no longer available. The Amos 3000 Super Widget (Amos Co., Clifton, NJ) is essentially identical to it, though, and could be substituted in this work if the following modifications were made . . . ." In other words, I recommended that he cite the name, but not the manufacturer of the original product (which had gone out of business), and instead give the name and address of the manufacturer of the substitute product, because both the product and manufacturer were not widely known.

Procedure descriptions

How detailed should authors get in their description of a procedure? Again, that depends on the nature of the work and the knowledge of the reader. In the previous example, the modifications to the Super Widget should be described in detail. If the procedure is one that all researchers in a given field use every day (moisten a soil sample, freeze dry a plant sample, etc.), you needn't use very detailed instructions, unless you have deviated from the normal practice. Authors who use a relatively new or rarely used procedure should describe it in detail or cite a reference where it is described in detailed. And, of course, if the paper describes a procedure for the first time, it should be completely detailed.

How to do all the above

If the information I have presented thus far seems clear, you are probably an experienced writer and may jump ahead to the next item in this chapter. If what I wrote seemed a little too general and you wish I had presented some specifics on *how* to write for readers, stay with me.

For many, "THE READER" is an amorphous mass that is about as difficult to get in touch with as a cloud. How can you write for readers you have not seen and know little about? Don't even try to write for THEM, write for a reader you know. Look down the hall of the building where you work or think about people you have met at a professional meeting. Probably somewhere in all your contacts you have met at least one person who reads the same journal you will be writing for. More likely you will know several. In a couple paragraphs I'll help you decide which one to select.

When I first started writing education news for the newspaper in Dubuque, IA, I kept a picture in my mind of a woman who regularly attended school board meetings. I figured she was a knowledgeable person, yet did not know all the information that I did. So I imagined she was sitting across a table from me, perhaps sipping from a cup of coffee. I wrote as though I was talking to her, and continued to use this technique until I had become comfortable enough with the subject and the audience that I no longer needed to use her as a crutch.

Then I left newspapers and started writing for the old *Crops and Soils Magazine*. That was quite a change; how was I to write for its readers? I used the same technique again; I figured that an extension specialist would be typical of the readers, and kept a mental picture of a particular extension person I knew, and wrote as though I was sitting with him in his pickup, explaining something I had just learned. After a while, I had developed my writing technique for that magazine's clientele to the point that I no longer had to use this device. In both cases (and many others as well) I would return to my mental image any time I would find myself in a bind as to how to present some particularly complicated new idea.

So if you want to write for readers of a particular journal, try to get a mental picture of someone from Category (ii) or (iii)—second paragraph of this column—who reads the particular journal you are writing for. Then write to him or her. If you have trouble getting started with this technique, actually insert the person's name in the first draft of your paper. Start out with, "Hey, Paul, ..." Further on in the paper, "The materials I used, Paul, were . . . ." Your conclusion could begin, "It seems to me, Paul, that this means . . . ." Once you have the first draft down, and perhaps your first revision of it, you can delete your friend's name, but remember the three Cs, and don't use that as an excuse to totally rewrite the paper into stuffy "scientific" verbiage.

If you are writing for something other than a scientific journal, consider who will be reading what you write. The readers of scholarly books and monographs will be similar to readers of a journal, but the readers of a high school textbook or a local newspaper will be substantially different from not only journal readers, but also from each other. The same technique applies, though. If you are writing a high school textbook, for example, think of someone in high school you know who is likely to take that class. The particular person doesn't matter, it could be a son, daughter, niece, nephew, grandson, granddaughter, or next-door neighbor. Whomever the person is, write to him or her.

If you are uncertain about a particular phrase or section of your paper, ask that person (if available), or another person who will read the material you are writing to let you know if he or she can understand what you have written. If that person has trouble with it, so will a substantial percentage of the other readers in that same category. Listen to their comments, weigh them for what they are worth, and heed them. Use the technique St. Paul described in one of his epistles, "Test everything and hold onto what is good."

I'll expand on the concept of writing for the reader in the next item and give you some specific questions that can help you write papers or speeches for a variety of people in a variety of situations.

* * *

What does your reader know and need? (part 2)

Last month's column opened the subject of considering the needs of the reader. There is much more to be said, so I'll devote this column and at least one more to it.

Last month I suggested that you ask yourself, "Who is the reader?" Readership will vary depending on whether you are writing for a journal, textbook, or for some other outlet. Merely knowing *who* the reader is, though, is not enough. This month I will present a little more about who the reader is, then move on to five additional questions to ask yourself before and while you write.

1. Who will read this?

You probably vary your speech depending on your audience. Suppose, for instance, I walked into your office and asked, "That experiment you ran yesterday; how did it turn out?"

After you have decided how you would answer that question for me, consider how would you answer if that same question were asked by:

- The graduate student who was supposed to run the experiment, but could not because he or she was home sick?
- The head of your department?

- A representative of the company that funds your research?
- Your spouse?
- Your 4-year-old child or grandchild when you get home tonight?
- Your next-door neighbor at a party this weekend?
- An environmentalist opposed to the product you are testing?
- A legislator who wants to know how to vote on a bill to ban the product?

The level of interest of these nine people (including myself) varies from almost none to intense. Their educational backgrounds vary as widely. On top of that are the level of interest and/or skepticism they bring along with their question. Some like your work and some dislike it. Still others (child, neighbor, etc.) may actually care little, if anything, about your work and are using it only as an opening to other conversation.

You may think that the people who read a scientific journal would not have such a high degree of variation, but you would be mistaken. Their variation is just as great, but in a different direction. In addition to what I wrote about individual readers last month, please remember that all readers of a scientific journal will be intelligent, but: (i) a few will know a great deal about you and your work, (ii) most will know nothing about you or your work, and (iii) a few will barely be able to read English.

Wise authors will vary their writing, based on the audience, so as to present something that is understandable to each audience.

2. Why will they read this?

Simply knowing *who* will read your paper is not enough. *Why* a person might read your paper is much more important to keep in mind. For example, a physician will read a scientific paper on the toxicology of a new chemical far differently than she will read an extension article on starting a garden in her back yard. The first she *needs* to know if she is to properly treat her patient; the second, she *wants* to know in order to find out how much work is involved and if she has the time and energy to take up gardening as a hobby. The place where you publish the information, as just demonstrated, will help screen the needs of readers for you, but there is more to it than that, which brings me to my third question.

3. What do they already know?

The answer to this depends as much on the subject as the reader. The physician in the previous example knows a great deal about toxicology, but little or nothing about gardening. A farmer would know a great deal about fertilizers and agricultural chemicals, and particularly would be well versed about their benefits as well as their risks. A typical city dweller may be aware of the risks, but not the benefits. Like you, other research scientists will know quite a bit about their field of study, but may not know much about your contribution to it.

4. What do they still need to know?

Just as Question 2 is the logical extension of Question 1, this question is the logical extension of the one that comes before it. The research scientist needs to know only a little about the basic context of your work, but much more about your results and discoveries. The farmer needs information on new chemicals, but not a long background on how fertilizers work. The physician needs little information on toxicology in general, but a great deal about the new chemical. For her backyard garden, though, she would need both substantial background information, a detailed explanation on how to establish and maintain that garden, and perhaps a bit of background information on benefits vs. risks, but not as detailed as you would give for someone else, because she is already well aware of the benefits vs. the risks of prescription pharmaceuticals.

5. Is there something I know that the reader needs to know?

There are two answers to this question. The first is obvious; if you have information on the cutting edge of research, let other scientists know it so they can further their own work. If you have information for an extension publication, by all means write about it and publish it in a place where the users of the information can get it.

The deeper answer is that many more people besides scientists and your usual extension-type clients need to know solid scientific information as well, but are not getting it. What do you do, for example, when you hear a network reporter or anchor give totally incorrect information about something in your area of expertise? Do you laugh at them or take steps to correct their error?

Thirty years, or even 20 years ago, most scientists would take the first route. Their reasoning, "I am a scientist, not a public relations person." In more recent years, though, more scientists are realizing that factual scientific information will not get out to the general public unless they take steps to get that information out. That's why ASA–CSSA–SSSA sponsors Congressional Fellows, has representatives on the CAST board of directors, write reports for CAST, and use other groups and devices for getting the truth out to those who need it. These group efforts are important, but so, too, are individual efforts. If you see a need to get some of your information out, consider taking steps to fill that need.

6. How can I best present the information to them?

There are two answers to this question; *where* to present it and *how* to present it. I've already dealt with the *where* when I said to place your information in a place where your intended audience will be most likely to get its hands on it. If you want to reach scientists with scientific information in their field, use the scholarly publications they read. The typical lay person, though, will not even know where to get a copy of a scientific journal, but will read a local newspaper or a regional or national magazine, and listen to and watch local and national electronic media.

The *where* to put your information is fairly easy, but quality of your presentation is more difficult. Even in the area of pure science, it is not enough to merely write a paper, you need to write a paper *that people will read.* Each month hundreds of papers in dozens of journals compete for reader attention. Few, if anyone reads all the papers in even a single journal, let alone every paper in every journal. And you have even more competition if you are trying to reach a broader audience than just other scientists. Regardless of the medium you use, your task is threefold; get their attention (shout, "Hey, people, here's something you need to know), whet their appetite, and deliver what you promised.

In a scientific paper you get their attention with a good title. Next, you whet their appetite—tell them *why* they should read your paper—with your abstract, tables, and figures; all of which are studied by typical readers before they decide whether to read the text. And finally, you deliver on what you promised by giving the readers the information they need *in a way they can understand it.*

I once saw a paper about the development of a common bean for the tropics. Such information would, of course, be of greatest interest and value to native scientists in places such as Columbia and parts of Brazil, Central Africa, and India. The second sentence of that paper read, "A bean ideotype for efficient N use would distribute N in time and space so that the leaf N (and therefore leaf photosynthetic capacity) was optimized with respect to reproductive N sinks."

Although I had been reading scientific literature for more than a quarter century, I had to read this sentence several times before I understood it. How many times would third-world scientists with poor English skills have to read it? More importantly, would they think the entire paper was written so obscurely that they could not interpret it, and stop reading at the end of that second sentence? In this case the readers would have been wrong. Aside from its second sentence, the paper was not badly written. I wonder how many got past that second sentence, though?

That sentence should have been written more like this: "Since N is limited in tropical soils, we need to develop bean cultivars that are good at both fixing atmospheric N and scavenging what N is in the soil, then apportioning that N as efficiently as possible between leaves and seeds."

[Crop scientists use the word *cultivar*, meaning *cultivated variety*.]

Don't bury your information either. In an earlier column (repeated in Chapter 2 in this book) I described a paper that began with an extensive literature review. The paper was rejected because "it obviously contained nothing new." The young graduate student, who did not want to have his first scientific paper rejected, wrote to the editor and said, "You didn't read the paper closely enough. All the research work I cited was done in Europe; my paper describes the first time this was done on the North American continent."

The editor countered, "It was not a matter of my not reading carefully enough, it was a matter of you not presenting it clearly enough. Rewrite your paper so that fact is clear and I'll take another look at it." The student did, and the editor not only read, but also accepted the revised paper.

This example demonstrates that writing so the reader can understand benefits the writer as well as the reader. I'll have more on that next month.

* * *

What does your reader know and need? (part 3)

A poorly written paper takes longer to publish and will be read by a smaller audience than a well-written paper. That's assuming, of course, that it gets published at all. I can't offer rigorous scientific proof for this statement, but think about the following scenario.

Time in review

You receive a manuscript from a journal editor for a peer review.

Can't do it today, you have a staff meeting and appointments with three students. Tomorrow you must prepare for a seminar you're giving the following morning. After you give that seminar, though, you have a class, another conference, and what appears to be an open hour at the end of the day. You decide to review the paper during that hour.

When the time comes you try to read the paper, but find poorly written gibberish. There seems to be a hint of something good, but you're not certain. Also there is something about the statistics that doesn't look quite right, but you're not certain because the paper is so difficult to read. "It's been a long day," you say to yourself. "Maybe I'm too tired to work on this right now." You set it aside, check your calendar for next week, and see four clear hours on Thursday morning.

Thursday comes, you try again and find the paper as difficult to read when you feel fresh as it was when you felt tired. You are tempted to throw it into the envelope with "REJECT" scrawled across it in big red letters, but realize that wouldn't be very professional. Besides, you've scheduled all morning to work on the manuscript, so you plow ahead. After two exhausting hours you check your calendar again and decide to take the manuscript home so you can work on it all day Saturday.

On Saturday, after you've mowed the lawn, taken the kids to soccer practice, and picked up a couple gallons of milk, you pour yourself a cup of coffee, settle into your study, and try again to review the paper.

After little headway during the next hour and a half, you realize you have two choices; recommend rejection or, in view of the time you have already invested in it, spend even more time to ask the questions and make the comments that might make it marginally acceptable.

Either way, two weeks have passed since you received the paper and promised a quick review. If it had been clearly understandable, with few scientific, technical, statistical, grammatical, or other problems, you might have been able to get the paper back to the journal editor after the first hour you spent with it.

Papers that require more-than-average time in review often need more time and work in subsequent processes (author rewriting, final review and scientific acceptance, and style and grammatical editing by the professional editing staff of the publisher).

Size of audience

Once a poorly written paper is published, many people may try to read it, but few besides the author will have the time, energy, or in some cases the ability to wade through. This can be unfortunate, as I mentioned in last month's column; a paper about a tropical bean cultivars contained information essential to the work of third-world tropical scientists, but written so obscurely in some parts as to scare off those scientists.

A better-written paper will be read by all the scientists in the author's field. A still better paper will also be read by scientists in related fields. As a paper gets better still, a few nonscientists will read it, too. Outstanding writing will be read by millions.

The astronomer Carl Sagan is not the ideal example because he often spoke and wrote outside his area of expertise. But if you can get past that, you have to admit he knew how to write about astronomy in a way non-astronomers could read and find fascinating. Millions of people, scientists and nonscientists alike, have read his magazine articles and watched his television programs. You can bet that *he* thought of *his* audience, because any article he wrote for a popular magazine or television show was vastly different from a paper he wrote for an astronomical journal.

Now it's our turn

The general public is at least as interested in food (where it comes from, how it's delivered, whether it's safe, etc.) as it is in astronomy. The same could be said for the public's interest in the use and misuse of soil. The general public listens to people who know how to present information on soil use and conservation and food and fiber development and production. The problem is that most of the people speaking to and writing for the general public about these subjects are not nearly as knowledgeable as is a typical member of our scientific societies.

As a consequence, the general public "knows" that today's scientific agriculture is not nearly as good for the soil, crops, or consumers as the type of agriculture practiced 100 years ago. The general public also "knows" that anything developed by a scientist could not possibly be natural, and therefore must be bad. Also, anything that has chemicals in it simply has to be bad. (I once asked a salesperson about the chemical makeup of a fertilizer, and was told, "Oh, it doesn't have chemicals, it's all natural.")

Wouldn't it be nice (some would say *essential*) if a knowledgeable scientist could write about crop and soil science and agriculture and food as well as Carl Sagan wrote about astronomy?

Choosing your medium

Without question, traditional scientific journals or scholarly books or extension outlets are usually the first and best ways for scientists to publish their information. In other cases, an electronic online journal is better. A quick e-mail or a message posted on your Web site might be preferable for still other sorts of information.

Some information, though, is not for scientists and requires speedy dissemination. For instance, when an emergency happens scientists often have valuable information for law-enforcement people, emergency crews, or the general public. This information must get out immediately, not a day later in a newspaper, a few weeks later in a magazine, or a few months later in a journal.

I can't help thinking about a situation in Madison, WI. The owner of a company that sells specialty gases arrived at work one Monday morning and noticed a canister of nitric oxide missing. "Why would anyone steal nitric oxide?" he asked himself, then gasped as he realized the most likely answer was that some high-school or college-age person confused it with nitrous oxide—laughing gas—and planned to use it to get high.

The gas distributor immediately got himself onto several radio and TV stations and explained that a couple of whiffs of nitric oxide would not make a person laugh and get high, it would kill them. Within a couple hours he received an anonymous telephone call from someone with a young-sounding voice: "I happened to be driving down a country road and saw your gas bottle lying in the ditch." [Sure he did.]

The distributor had no desire to prosecute; he was grateful that the burglar heard his message before that first deadly whiff, and told him where to retrieve the canister.

I am not saying scientists should stop publishing in paper or on-line scientific journals, scholarly books, or extension outlets. But those who do that well and feel comfortable writing or talking to a more general audience, should consider exercising their talent. Write an occasional popular article or op-ed piece, make a call to news media (local or national). Or call your governor or legislator or congressional representative or someone in the national administration. Few scientists are capable of doing this and most should not even try. But *some* scientist (maybe it's you) should if we ever hope to do something about the scientific illiteracy of the American public.

* * *

To sum up those three columns, once you are aware of the needs of a particular readership, the sort of information those readers already possess, and consider the sort of information you possess, you can make up your mind as to whether it would be a waste or a good use of your time to write something for that audience. This is just as true for those in technical or nonscientific fields as it is for scientists.

I went through exactly that thought process as I set out to decide whether to create this book out of the nearly 100 columns I had published during a time that stretched close to a decade. Was there need for this information in a more permanent and easily accessible format? Based on comments I received from members of ASA–CSSA–SSSA, the answer to that question appeared to be yes. Was I the best person to compile the original columns, update them, and arrange them into the logical sequence of this book? I couldn't think of anyone who would be any more qualified. And especially I couldn't think of anyone who had the insights and background into why the columns were written in the first place so as to be able to update them in a way that they truly were updated.

You are reading the fruits of that mental exercise.

I urge you to perform a similar exercise as you consider the knowledge you have and the needs of the world, or some segment of the world. If someone needs information you have, and if you have a better ability than others to deliver that information in an understandable way, then you have two good reasons to deliver your information to those who need it.

Chapter 2

Beginning to write

Okay, you've read Chapter 1. You know you have information. You also know who needs that information and why. That's the answer to the question of whether you should strive to get that information written, submitted, and published. Now, where and how do you begin to actually write down that information for them?

Some people are deathly afraid to face an empty sheet of paper or an empty computer screen. I have never had that particular fear myself. In fact, I can't wait to fill those empty spaces with something that is either entertaining or informative or, preferably, both. But fearful types ask themselves questions such as: "What if I don't start correctly and have to begin all over again?" "What if nobody likes what I write." "What should I write first?" "Where do I begin?"

I have a suspicion, based on a few personal observations, that much of this fear comes from traumatic happenings early in a person's life. In many of these cases the person wrote something that was either poorly received or even ripped apart because of its spelling or grammar, with no acknowledgment whatsoever of its contents.

I have faced this from time to time myself:

Writer: "What do you think of it?"

Critic: "You misspelled this word."

Writer: "Well, okay. No big deal. That can easily be corrected. What about the contents of the paper, though? Do my ideas make sense?

Critic: "You also misspelled this word, and this word, and three words on this final page. And this sentence in the middle of this page isn't a complete sentence."

There is little an individual can do when confronted by a critic of this sort except ignore him or her and find someone who is able to evaluate the creation on its merits instead of its peripherals. I have developed a thick enough skin by now that I usually can do this.

For those who have not yet developed this ability, I wrote out the words of my philosophy, which I have posted on the home page of my Web site, and a few other places:

"The whole message I'm trying to convey is that scientists (or anyone who feels uncomfortable writing) do not have to be afraid to write just because they can't define a gerund and failed to memorize every grammar rule Miss Wormwood taught in third grade. If such people would loosen up and write more, the process of writing will begin to come more natural for them and they will eventually become better writers. Many of the letters I have received from scientists through the years have told me this is exactly what has happened to them."

I addressed this subject in a slightly different way in the following column, published in July 1996.

Bad writing has its place

During the ASA–CSSA–SSSA Annual Meetings in St. Louis last fall I had an interesting conversation with Abdouh M. Yacoubi, a consulting agronomist from Virginia.

He began by holding up his program and saying, "There are 2,100 papers in this book and half of them are terrible."

I rolled my eyes in tentative agreement.

"But this is good," he quickly added. "Americans allow bad papers to be presented at their meetings and in their journals; Europeans don't. In Europe a person isn't allowed to publish until he can do it perfectly. As a consequence, everyone is afraid to publish and nothing gets published."

"Do they allow their children to crawl before they walk?" I asked.

He nodded. "It's exactly the same thing. I presented my first professional paper at the 1974 ASA annual meetings and it was terrible. I presented my second paper in 1978 and it was terrible also. The paper I presented in 1982 was fairly good, and from 1986 my papers have been good. Had I not been allowed to learn from my mistakes, I never would have been able to develop my career to where it is now."

I nodded, "I see your point."

He continued, "In any group of scientific papers, perhaps 10% are worthless, another 20% are bad, 40% are all right, 20% are good-to-excellent, and 10% are outstanding. If you begin with a field of zero papers, though, the 10% that are outstanding will total zero papers."

I asked him, "What can be done to loosen things up in Europe?"

"Europe isn't unique," he replied. "There's a similar problem in Asia, where it's considered impolite and insulting to criticize another's work. As a consequence, nothing gets challenged and science can't develop."

The case for bad writing

Since that day last fall I've often thought about Dr. Yacoubi's words. While I don't particularly like to read bad writing and would never advocate it, bad writing is preferable to no writing at all; particularly for people who are just getting started in their professional careers.

The trick, as Dr. Yacoubi stressed, is to *begin.* Regardless of how good you think your first paper is as you are writing it, your second will probably be better. By the time most people get their fifth or sixth paper published, they may begin to wonder if there is some way to gather up every copy of the journal that contained that first paper and burn them all so that the colleagues who know them now will not accidentally stumble on it and read it. But face it, if it weren't for the fact that you wrote and published that first and second and third paper, your fifth and 10th and 15th paper could not be as good as they are.

Eliminate your fear

Some people have told me, probably not entirely in jest, that they are afraid to write to me because I might see something bad in their writing. The opposite is more likely the case.

A recent letter began: "Dear Mr. Luellen: Please excuse my poor grammar, sentence structure, and punctuation. By my writing you wouldn't know that English is my first language." After all that groveling, the author of this letter asked his questions, then concluded, "If I haven't expressed myself well enough to understand, please let me know. If you have the time I would be most appreciative and interested in your response. Thank you for your time."

I felt terrible and wondered what I had ever done to intimidate this poor fellow so badly. I began my reply to him with a profuse apology for whatever it was I had done that had caused him to feel so inferior. His response? No, I had never done anything to intimidate him, he was merely ashamed of his writing.

He had no reason to be ashamed. His writing wasn't so bad. In fact, I've seen worse from people who think they are good writers.

If fear is keeping you from writing, put aside that fear and write anyway. You can learn something whatever and whenever you write, that will help you write better the next time. Note the word *can* in that previous sentence. It's not automatic, people learn from their mistakes only if they are open to it and willing. If you believe you never make mistakes, I feel sorry for you; you will never be better than you are today.

Critique yourself

Days, weeks, or months after I have written something (even if I have already mailed it off or seen it in print) I read it again and ask myself if there might have been a better way to present the information. I do this in private, and try to be brutally honest with myself. In my earliest days I would often need to say, "This is worthless trash." A few years later I seldom said those words, and usually could say, "This isn't bad, but there is room for improvement." Today I cannot remember the last time I had to say the first of those previous two sentences and usually say the second. Increasingly, now, I am able to say, "This was a damned good piece of writing. I'm proud of it."

During the review of your next scientific paper, heed the comments from scientific reviewers and editors, and your publisher's professional editing staff. Most of those comments will be fair and correct. Keep in mind, as you read the comments, words written a couple thousand years ago by a rabbi named Paul, "Test everything and hold onto what is good."

After your paper is published, listen in the same way to your colleagues who read and comment on it. Also, at some point, read it again yourself, critically.

The only way you can ever get better is by admitting your mistakes and learning from them. As soon as you begin to believe your writing is as good as it ever can possibly be, you will be correct. Some may be satisfied with that, but I prefer the way the architect Frank Lloyd Wright put it when asked which of his buildings he considered his greatest achievement: "The next one."

* * *

After publishing that column I received a few critical letters from people who took Dr. Yacoubi's words literally: "How can he say no scientific papers are published in Europe and Asia? There is lots of good science published there." I doubt he meant his words to be taken literally, and I know I neither heard, nor attempted to present them in that light.

Far be it from me to put my own meaning to his words, but *my* interpretation of them is, "Compared with what might be published if journals were not so rigid, there is very little science getting published in Europe and Asia."

The same can be said, to a lesser extent, in even the USA. If there is one complaint I have heard more than any others, that complaint is journals are too rigid. Their reviewers set the bar (as in pole-vaulting) too high. Only the tried-and-true boring material makes it through to publishing; the creative, exciting, innovative material does not.

There is, of course, another side to that question, which has been clearly and eloquently stated by those who argue for keeping that bar high so only truly the best gets through.

If you stop and think about it, generally speaking, the most prestigious journals are those with the lowest acceptance rate. The acceptance rate for the journal, *Science*, was, the last time I checked, something like 0.001%. Think about that the next time you grumble because the journal you submit to has an acceptance rate of "only" 60%.

Which leads logically to the next column, published in February 1993. The title was startling to some when it was first presented, and may be startling to some of you here. After all, in this chapter titled "Beginning to write," why have something so negative so close to the beginning? Quite simply because rejection is as natural a part of writing as is acceptance. And rejection, as you will read, is not the end of the road for a paper, but merely one step along the road toward eventual publishing of it.

After your paper is rejected

A graduate student wrote a manuscript listing his faculty advisor as co-author, submitted it to a journal I managed without allowing the more experienced scientist a chance to see the manuscript, and left the country.

The manuscript contained a number of flaws, not the least of which was the lack of an abstract. I could not find a forwarding address for the student, so I returned the manuscript to the co-author, pointing out the paper's shortcomings.

The co-author called me. "I certainly agree with all you wrote; the manuscript was nowhere near ready to submit. Had the student shown it to me first I certainly would have told him of the shortcomings and worked with him to get it correct before submitting it. What can I do now; does your rejection mean I can never again submit it to this journal?"

Judging by my correspondence and phone calls, many authors share this person's concern. They seem to think that once a paper is rejected by a journal, it can never again be submitted to that journal. Consider the following letter from an author to a technical editor who rejected a paper:

"I would like to thank you for the papers you sent to me about journal writing. Using the information in them, and the comments from the reviewers of our paper, I revised the manuscript and submitted it to the *Canadian Journal of Plant Science*. The manuscript was accepted with only minor revisions necessary, due in great measure to your advice and that of the original reviewers."

The technical editor said to me, "I didn't specifically invite the author to resubmit the paper to us, which was probably a mistake."

When editorial board members suggest ways for an author to improve a paper, it is not so the author will take those suggestions, improve the paper, and submit it to another journal. Your associated Societies are as anxious as are its authors to publish good material in *our* journals.

A release or rejection of a particular manuscript is just that; it is not a rejection of the person, not a rejection of the research project, and not a rejection of anything else.

Here are some specific steps an author should take when he or she receives a letter of release.

Contact the person who rejected the paper

First, read the rejection letter carefully to see if you missed something or if, perhaps, the reviewers missed something. If you find the latter, contact the person who rejected the paper and discuss the matter.

A good example of this happened with the first paper of a now prominent scientist. In his literature review, the young graduate student cited a large number of other papers on the same subject. His paper was rejected on the basis of that large review because, "Obviously you have nothing new here."

The young scientist replied, "You didn't read my review carefully enough. Every one of those papers I cited were from Europe. My paper reports the first time this work was done in North America."

The technical editor replied. "Rewrite the paper so you make that more clear in the introduction and I'll take another look at your paper."

The young student modified the paper with that mind, resubmitted it, and had it quickly accepted.

I know of one now-retired scientist who regularly made it a practice to call the editor who rejected his paper and argue about the release; sometimes, it seemed, even when he knew the reasons for the rejection were valid. I guess he figured there was no harm in trying. (It was not a good way to keep himself on good terms with journal editors, though; see my comments about "the halo effect" further down in this chapter and also in the next chapter.)

Appeal the rejection

If you cannot get the person who rejected the paper to agree with you, but you are certain you are correct, appeal the rejection. The appeal procedure varies slightly from journal to journal, but is spelled out in the style manual of most publishers. The appeal will be on the rejection procedure, not of your work. In other words, appeals are based on the paper *as it was submitted.*

For instance, if the reviewers said there was a flaw in your procedures or statistics or replication or whatever, and you can demonstrate that those reviewers were incorrect, your appeal will be successful and your paper will be reinstated.

Rewrite the paper and resubmit it

If the paper was rejected because the reviewers wanted another year of data; correction of a major error in the design of the experiment; or some other major restructuring of the work, the paper, or both (and you agree the reviewers were probably correct), re-do the experiment so you get that additional data, or recalculate the data using a different statistical procedure, or correct whatever else is missing or incorrect, revise your paper, and resubmit it.

Rejection not the end

To sum up, the release of a paper does not spell the end of that paper.

If you agree the release of your paper was for a good reason, correct the paper and resubmit it. If you disagree, appeal the rejection informally first and then formally if necessary.

The goal of both authors and publishers is to publish the finest possible scientific papers. Release of one particular manuscript is nothing more than one step on the path toward that goal.

* * *

I hope the first two essays in this chapter have dispelled your fears, and now you are neither afraid to begin, nor willing to be discouraged by the occasional inevitable rejection. You are, in fact, ready to begin writing the best paper you can.

Where and how, specifically do you begin? One good place is with careful consideration of the way your brain is constructed. That brain of yours has three components with three specific functions. The best writing happens when writers know how to call upon the unique qualities of all three of those components. Where can you learn how to do that? In the next column, which was first published in December 1993:

Use all of your brain when you write

The right side of your brain is the creative side; the left is the logical or linear side. Nobody knows where the subconscious mind is, but its forte is synthesizing things. People have known this for years, but it was only recently that writing experts, psychiatrists, and psychologists have devised ways that allow people to make use of all three parts of the brain when they write.

Going into the techniques in depth would take an entire book (at least). But I can give you enough practical information in this column to enable you to vastly improve your writing quality.

How to get started

What are you going to write about? At first glance the answer to that question is simple; "I'm going to write about my research project." Take another look at the question, though, because the quality of your entire paper, whether and how quickly it will be published, and the impact it will have on the scientific community and perhaps the rest of the world all hinge on the *proper* answer to that seemingly simple question.

Your scientific paper will be not about your research project, but about the *finding(s)* of that project. A good working title for any scientific paper could be, "We discovered something nobody else has ever known." [Engineers, physicians, and technicians can come up with something similar.] When you use a working title like that, it will help you focus on the important thing(s) and keep the peripheral to a minimum.

A better way to outline

Some people are enamored by outlines. I don't want to say outlines are useless, but they make use of only the left side of your brain and shut out your center of creativity.

A replacement for the outline that uses both sides of the brain is called clustering. If you would like to try clustering sometime, begin by taking a large piece of paper—say wrapping paper or poster paper or something of that sort—or use a chalkboard in an unused classroom. Write your main topic in the middle of it.

Find the main topic

Before you can write down your main topic, you have to figure out what it is; no small task. Think about it seriously. Your topic will not be, "We researched a question for three years and finally have an answer." Every scientific paper in the world could use that. Your main topic is, "We developed a wheat that will grow with one-third the water everyone else believed was the absolute minimum." Or maybe your topic is, "Everybody else was wrong; we found the actual genesis of this soil."

You should be able to state your main topic in fewer than 25 words—preferably far fewer. The imaginary topics just given were stated in 19, and 12 words. If you can't state your main topic in one short sentence, you haven't defined your main topic. Keep working until you find it. (As a newspaper editor once said to me, "Even the largest and most important subjects in the world can be stated in five or fewer words. For example, World War III started today.")

Brainstorm your ideas

Once your main topic is written in the center of this large piece of paper or chalkboard, cluster all the related topics around it. You do this by writing down a word or two or a short phrase to represent each idea and thought connected with the main topic. Don't do a lot of thinking about it, just let the ideas and words flow. If possible, write the ideas that are related to each other—those that belong in Materials, Results, etc.—somewhat close together (that's why you started in the center of the paper or chalkboard). But leave plenty of space, too, so you can add more words and ideas, and, still later, draw lines linking them to each other and to the main topic.

If you have never done this sort of thinking, you probably will have difficulty with it at first. This is normal, and brought about because the left side of your brain is reluctant to give up its dominant place and control of the unruly right side. Speak to it gently. (I know this sounds silly, but it works.) Say to your left brain, "Give the right side its chance. It will be okay; I'll call on you later to help me make sense out of it all."

It is important not to censor or edit yourself at this stage of the project. Remember, the creative right side of your brain thinks in nonlinear and nonlogical terms. Let the thoughts flow as incoherently as they might; jot down everything that comes to your mind with only a minimal amount of attempting to organize it. Don't erase a thing. Another word for this process is brainstorming; let your brain pour forth related and seemingly unrelated ideas like a storm.

If several people are writing a paper together, all might stand around the paper or chalkboard calling out ideas, or writing them down themselves. This is not a time to interfere with one another; this is a time for an entirely free flow of ideas. Nobody should ever utter the words, "No, we don't want to get into that in this paper."

After 15, 20, or perhaps 30 minutes or more of this brainstorming, when you are certain you have written down every possible word and idea, step away and take a few deep breaths. It would be even better if you would leave the room for a cup of coffee or a soft drink. If you can spare a couple days, so much the better. However long this break is, don't think about the writing project during the time.

What you are doing with this break is allowing your subconscious mind to continue to work on the problem. If your conscious mind is also working on it, your subconscious mind will be occupied with absorbing information from the conscious mind and not have time to do any sorting and synthesizing or what it had picked up earlier. After the break, come back, look at your paper or chalkboard again without too much deep thought, and write down any more ideas, thoughts, and words that come to the fore. (Surprisingly, there will be more; supplied by your subconscious.) Listing these might take another 5 to 15 minutes.

Logical synthesis

The right side of your brain and your subconscious mind have done their parts for now. (You will be calling on them again later.) Now allow the left side, the logical and integrative side of your brain, to take control for a while. Step back and study your paper or chalkboard.

Are some of the words and short phrases obviously linked to each other? Circle them and, if they are separated, connect them with lines. Use additional lines to link them to other clusters of words and phrases. Use circled numbers or letters to help you sort out how and where they fit together under your main topic. You will find yourself thinking things like, "These items here are the basis for the Materials and Methods section, and these other ideas can go here with them. This part, I don't want to put into this or any paper. These ideas belong in an entirely separate paper. Look, here's my conclusion, plain as day."

How long all of this sorting out will take will depend, to a great extent, on the complexity of your project and the length of the paper(s) it will take to report it.

Writing the first drafts

As soon as your thoughts are organized, you are ready to begin writing the paper. If at all possible, do it immediately after the organizational step just described.

It is time to call on the right side of your brain again. Plan to give yourself enough time to write your entire paper, or at least a substantial section of it, without stopping: ("I'm not getting out of this chair until I have all of Chapter 3 written down.")

Keep the guideline sheet you developed from your large paper or chalkboard jottings in mind and go as quickly as you can. Do not think about anything too carefully, and pause for nothing. Race all the way through the paper or the section you are working on, not even pausing to correct the obvious typos (unless you absolutely cannot tell what the word is supposed to mean).

Once you are done, take a break—how long or short depends on your personal needs and the urgency of your deadline—but make it as long a break as you can. This break will give your subconscious mind time to ascertain whether you left anything out. After your break, start back through the paper to see what you have. This shifts you back to your left brain, with help from your subconscious. (It helps immensely to be working on a computer or word processor at this point, although the following can be done with a pencil on typewritten or pencil copy.) Read what you have written, asking yourself such questions as:

- **Is everything there that ought to be?** If not, add it now or make a note to look up the needed details and add them later.
- **Are any paragraphs out of order?** If so, move them. Are sentences out of order within paragraphs? Move them.
- **Is there some incredibly bad wording?** Correct it.
- **Are sentences too long and wordy?** Shorten them.
- **Are some sentences repetitious of others?** Select the best and delete the rest, or select the best parts and combine them into one, then delete the rest.

In short, make use of all the tips in this book.

Unless you are completely exhausted, once you have gone through the paper and made your changes, immediately go back to the beginning and start through it again, and again. (Unless the paper is exceptionally long or I am extremely tired, at this stage I generally try for six to seven times through the paper at one sitting, making changes each time). Then give it a rest. (At this point I print out my first hard copy—what I jokingly call my first draft.)

Your first break after this first writing session should be at least 10 minutes long; a day or two or even a week or two would be better, if you can spare that much time. Try not to think about your paper during this break. This allows your subconscious mind time to continue to work on the project.

Come back after your break, take another look at the paper, and revise some more. (I begin this stage by working with pencil on the "first draft" hard copy. Then, as I transfer these changes to the computer, I make more corrections, changes, and revisions.) After I have read and corrected it once on the computer screen, I read it three or four more times, making additional revisions each time. When I am reasonably satisfied, exhausted, or both, I print out a hard copy (my "second draft").

If the deadline is at hand, this "second draft" is the version I use. If more time is available, I take another break to give my subconscious mind another crack at it, and begin another round of reading and revising. I repeat these steps as many times as possible within the available time.

When you have completed your paper (in a sense, it will never be completed, only abandoned), you are ready to write your abstract. This is both a logical and a creative process. Let your left brain work at it first; read the whole paper and jot down the most important parts. Give your right brain first crack at condensing these parts into the abstract by writing it as quickly as you can with as little thought as possible. Then go back, read what you have written, and make needed revisions. (I usually rewrite an abstract no more than two or three times before I print it out, and may use the first, or at least the second printout as my final version.)

Finally, find the key words in your abstract and make them into your title. (As with the abstract, another left, then right, then left brain function. You will probably find the title will resemble, but not necessarily be identical to, the main topic phrase you wrote down in the center of your paper or chalkboard when you first brainstormed this paper.

How long does all this take?

Although it sounds like this method of revision after revision will take a great deal of time, this technique should take no longer than any other writing technique. Basically what you are doing is starting with a better package, then making most of the revisions on it before the reviewers get a chance to ask you for them. Even beginning writers who have used versions of this technique have had papers accepted with few or no additional changes.

But I like to write my abstract first

I've known more than a few people who say they prefer to write the abstract first: It helps them to have a better idea of what they want to ultimately write about in the full paper.

If that's the way you like to write your paper, fine. I'm not dictating any one way or even arguing that the technique presented here is the best for everyone. But, for those who say it is better to write the abstract first, I have two comments.

First: You can write your abstract using many of the same techniques outlined here.

Second: Realize that when you write your abstract first, it becomes a sort of an outline for the rest of your paper. This, then, is better than a standard outline because, as mentioned above, outlines are written, for the most part, with only the left part of your brain.

Not just for scientific papers

Most of the things I write are far less complicated than a scientific paper, so I do not always need to use the large-paper-for-brainstorming-ideas part of the technique described here. When I begin to write one of these columns, for instance, I have a reasonably clear idea in mind of the subject and at least a rough idea of what I want to write about it.

For this particular column, for instance, my topic subject was, "How to use all parts of your brain when you write." I have read many books and articles on the subject, refined the technique during the past 25 years, and regularly use it myself. So I did not need to do much additional research. Instead, I sat started hammering on the computer keyboard.

After the first batch of words were on the screen, I paused, looked at what I had, and found it was not bad. I made three or four passes through it, stored it in the machine's memory, and turned to another project. Two weeks later I took another look at this column; made about three passes through, revising each time, and stored it again. About three weeks later I took a third look at it—only time for two passes through it that time—and made some substantial changes, including the addition of the sentence you are reading right now. The reason why I made such substantial changes was because the preceding interval included a vacation in the Southeast for sightseeing and visiting, as I sometimes describe them, my granddaughter and her parents. My conscious mind was not thinking about the paper at all during this vacation, which gave my subconscious mind plenty of time and freedom to work on it.

The next step for this paper was a review by the other managing editors in ASA–CSSA–SSSA Headquarters Office (my peer review committee). After receiving their comments I'll decide which of them I wanted to incorporate, added them in, made a few more adjustments, then let the paper sit a few days longer. Finally I went through it one (or two or three) last time(s), make a few more changes, changed the spacing from double to single, copied it onto a disk, and gave it to the editor of *Agronomy News* for his editing and publishing of it.

[It is no longer 1993, it is 2000. I had looked at this column about a year ago when I was getting ready to publish it in another place, and did some polishing of several of its sentences at that time. Then, a year later, I again revisited it as I prepared to put it into this book. I saw several areas that needed updating and revising. One section was added under the subhead "But I like to write my abstract first." That was in response to an article I read a few months earlier saying all papers *must* be written *only* after the abstract is completed. I knew I had to add information about that here, if nothing else so I could express my abhorrence for words such as *must* and *always*, or even *should* in any book of this sort.]

If you would ask me how long it took to write this column, I would have to honestly say "my entire life." As for total elapsed time spent in front of the computer actually writing and revising, I did not keep a stopwatch, but I would guess no more than 8 to 12 hours were devoted to the original column before I gave it to the managing editors for review and no more than 4 to 6 additional hours from the time of their reviews to handing the disk to the editor of *Agronomy News*. (The additions to it as it was compiled into this book took, perhaps, an additional couple hours.)

This is about the same amount of time, perhaps less than it used to take me to write before I learned this technique. (I freely admit the ability to revise and revise on a computer screen without having to retype one or more entire pages each time has made this technique possible for me.) The quality of my writing is definitely better than before I had a computer that allowed me to fully use the technique.

* * *

As mentioned within it, if you compare version of the column you just read with the version published in 1993, and a second version published early in 2000 you will see changes both times. The version sent out in early 2000, as a premium gift for persons who subscribed to my newsletter, had quite a few of its sentences polished, compared with the original. This version retained that polishing, plus had some additional polishing, the addition of several paragraphs mentioned within the column, and a couple other less-extensive additions. All of this merely provides evidence of what was mentioned within this column, that a work of art is never finished, merely abandoned.

* * *

Now I need to provide something else for you to keep in mind, particularly if you are a practitioner of scientific writing. That is a reminder of the fact that your paper will be around for a long time. The following information was first published in September 1997:

Are you writing for the ages?

One of the persons present when Abraham Lincoln died supposedly said, "Now he belongs to the ages." Those words also could be said of scientific writing, but are you writing with that thought in mind?

When modern is not

Modern is one of the most ephemeral words in the language. The only word I can think of that might be more ephemeral is *new*. I always get a chuckle when I thumb through a 30- or 50- or 75-year old (or older) publication, see a picture of a tractor or truck or aircraft or typewriter or whatever, and read the caption, "A modern . . . ."

Early in my writing career I decided I didn't want someone in the distant future getting a similar chuckle while reading something I had written, so I decided never to use the word *modern* in any of my writing (except as I'm using it in this month's column).

Avoiding the word *modern* is not as difficult as you might imagine. Here's a line from a manuscript I edited recently, "Modern corn planters have the ability to . . . ."

In most cases an author could simply drop the troublesome word without any substitute: "Corn planters have the ability to . . . ."

If, for some reason, you definitely need to distinguish the currently available models from those of an earlier vintage, use, "Present-day corn planters . . ." or "Today's corn planters . . ." or "The current generation of corn planters . . . ." If I needed to use a phrase such as one of those, I would be inclined to soften it by moving it inside the sentence, "The corn planters available today . . ." or "Corn planters currently on the market have . . . ."

The siren is calling

Like the mythical mermaid, science is constantly luring its unsuspecting practitioners to write: "Back in the old days when we didn't know much about [name the subject], scientists believed [name the incorrect earlier belief]. Now that modern science has discovered the real nature [cause] of this [subject], we can definitively state that the entire and complete answer to that dilemma is [the current state of the knowledge]."

Unless sentences of that sort are used in a review paper of a subject that has been well defined for a large number of years, I can just about guarantee that anything written with that much hubris will one day come back and bite its author.

In-vogue phrases

Fortunately, I see few instances of scientists using current in-vogue popular phrases. (You wouldn't use them at all in a scientific paper, of course. Avoid the temptation to use them in a more popularized paper as well.) For instance, I have never seen (and hope I never do) a scientific or popularized scientific paper containing the line, "Show me the money." Within a few months (maybe even by the time you read this) that phrase will be as obsolete as, "Missed it by that much," and "Would you believe . . . ?" For those who do not remember, both of those phrases were used as frequently in their day as "Show me the money" was when I wrote these words. But they are meaningless to today's readers.

I'm also concerned some of the terms that have recently come into vogue in scientific writing will one day seem dated and obscure as those popular phrases. *Paradigm shift* is one such term. Five years ago nobody heard of it, now virtually everyone uses it. Will this continue? I don't know. And because I do not know, I do not use it in anything I know will be around for a while.

I was concerned about *deleterious* when it first appeared, but it seems to have established itself and may have some staying power. I have not begun to use it myself, though, because *harmful* is such an admirable alternative.

As I've written elsewhere, *parameter* was once an excellent word, but it has been misused to the point that it is meaningless. I believe it will drop out of the picture altogether in the near future, along with *facility*, which also succumbed to rampant misuse.

What we can do

No single person, or even no professional organization of 12,000 or so persons, can hold back the evolution of our language. But we who write for the ages need to keep the ongoing evolution of our language in mind and try to write in such a way that our words, and the ideas behind them, will at least be understood, if not treasured, by the future generations who will read them.

* * *

Okay, you have used all your brain, you have written a piece that ought to stand the test of time, where do you go from here? You make the paper as perfect as you can *before* you submit it to a journal. "Wait a minute," you ask. "Can't we take care of details during the review process?" Of course. But the better the paper you send for the review, the fewer problems it will have in that review. Here are a few more words I wrote about the subject and published in September 1999:

DO sweat the small stuff

An envelope arrived here late last year addressed to Jay Poster, editor of what was then called *Agronomy News*. What was wrong with that? It contained a submission to *Crop Science*.

Of course, Mr. Poster carried the envelope down the hall to me and I forwarded it to a technical editor of *Crop Science* (to whom it should have been sent in the first place). I was tempted to return the paper to the author, though, because:

- the abstract had a large number of style problems including far too long a length, lack of an introductory statement, and no botanical and chemical names
- the manuscript was not on line-numbered paper and its pages were not numbered
- the manuscript contained a list of *additional index words*, which we have not used in any of our journals since 1988.

In short, the author sent an improperly prepared manuscript to the wrong person at the wrong publication at the wrong address. How important are each of the errors?

Of course, mailing the manuscript to the incorrect address could be devastating. Your Headquarters employees see that a manuscript that arrives here will get where it is suppose to go. I have no idea whether other publishers do the same. I remember once forwarding an incorrectly addressed manuscript to the American Society of Agricultural Engineers, but I cannot remember ever receiving a man-uscript forwarded from them. (I'm hoping it was because none were mistakenly sent there.)

As for the other errors on the submission:

- Inclusion of additional index words is minor, those words can easily be deleted.
- Lack of page numbers could be devastating if something happens during editing or review to mix up the pages (even something as simple as the manuscript accidentally getting knocked off a desk).
- Lack of line numbers forces the reviewer to find another way to identify items instead of using, "page 10, line 7."

Sometime during the review process an author will need to rewrite that improperly prepared manuscript anyway, so it makes sense—and will speed the review—to get it right before you submit it. I had a strong suspicion the paper mentioned above also contained one or more major scientific flaws, which brings me to an even more important reason why authors should try to get everything right before they submit their papers, that is the halo effect.

Halo effect

Halo effect is the name for the aura or appearance that accompanies anything or anyone, and is particularly evident at first impressions. If you or the things you produce look competent and professional, people will assume that you and they are. If you or the things you produce look sloppy and clueless, people will assume they are that also, until you prove otherwise.

When I submit something, I try to do so in a way that the editors will look at it and think, "Whoever submitted this knows what he's doing."

A poor submission can instill exactly the opposite thoughts in the person who receives it. For instance, if you prepare a manuscript in the proper format for one journal and then send it to a different journal, you invite the editors of the journal that receives it to form one of the following conclusions before they begin to read your paper:

- You don't know enough about our journal to know its format requirements.
- Or, you first sent your paper to that other journal and it rejected the paper, and after that rejection you did not think enough of us (or possibly your paper's chances with us) to convert it to our format.

Correct or not, are those the thoughts you want editors and reviewers to be thinking as they review your paper?

Find out to whom a manuscript should be sent, and address your manuscript to that person. Just as you probably do not like to receive a letter that starts out, "Dear Occupant," editors are not particularly fond of letters beginning with, "Dear Editor."

But even those are better than incorrect names. Certainly, editors understand and forgive such errors the first few months they are on the job, but I received one the other day addressed to "Steve Breth, Editor, *Crops and Soils Magazine*. I replaced Steve Breth when I came to ASA in 1969 to edit that magazine, and ASA discontinued the magazine in 1988. So the sender of that letter used an address that was 30 years out of date to a magazine that no longer exists. What would your thoughts be of a person who made such gross errors in something sent to you?

Heed the requirements

Every issue of every journal has a masthead containing the basic information on how and where and what to submit to that journal. Most journals also publish, from time to time, an "Instructions to Authors" or "Suggestions for Contributors" that contain specific and detailed information on how to prepare the paper and where and how to submit it. In addition, all ASA–CSSA–SSSA members have been sent a copy of those organizations' *Publications Handbook and Style Manual* (and also have access to its electronic version on those societies' Web pages). This book gives even more detailed instructions on how to prepare a paper for submission. If you need a copy of the style manual of any organization or publication, contact the appropriate book order department.

Please read and heed the information supplied in the above documents before you submit your next paper. It will certainly speed the review of your paper (compared with the time it takes to review papers with errors like those mentioned in this column) and it might even influence whether your paper is accepted for publication.

* * *

Speaking of the ASA–CSSA–SSSA style manual, this column from July 1994, asked and answered a question I had frequently heard:

Who writes that Style Manual, anyway?

A manuscript came back the other day with the following note, "Luellen may want to use 'genotype-by-environment' instead of 'genotype × environment,' but I prefer '×.'"

The author of the letter was correct in his terminology and incorrect in his reading of my opinion of it. The Boards of Directors of the Societies determined in 1990 that a *times* sign is not a synonym of, and thus could not substitute for, the word *by*. Thus a plot will be described as measuring 2 by 4 m, not 2 × 4 m. The term "genotype-by-environment" reaction, though, has been abbreviated as G × E for so long that the boards decided to allow its continued use as an exception to the rule.

There have been a few times when *I wished* I had the power to dictate what can and cannot be done in our journals. I don't have that power, though, and I realize I shouldn't. The journals are not mine; nor do they belong to any other single person. They belong to the Societies.

If I am not the person who dictates style rules, who does? There is no one person. Instead, various groups of scientists and others on nomenclature and style committees recommend style changes. The editorial boards of one or more journals consider the requested change and, if they approve, make their recommendation to the Societies' Editorial Policy Coordinating Committee. This committee decides whether the proposed change would be good for all our publications and, if so, makes the recommendation to the Boards of Directors of the three societies. The Boards of Directors have the final say.

I am a member of the Headquarters office editorial team, the editorial boards of three journals, and the Editorial Policy Coordinating Committee. As such, I have the right to take part in some of the discussions leading to the formulation of style rules. In some instances I take an active role in these discussions. Most of the time I choose not to take part: I do not believe it is the place for a nonscientist employee like myself to tell scientists the best way to present scientific information in publications of their professional society.

For instance, in the discussions that led up to the promulgation of the rule not to allow use of the multiplication sign to mean "by" (mentioned at the beginning of this paper), I chose not to take a side. My position was, "You scientists decide what is clearest to you, and I will go along with that decision."

Even though I don't necessarily take part in all discussions, I occasionally ask the questions that begin those discussions. For instance, I once asked a number of scientists whether the Munsell number alone tells all a scientist needs to know about a color. I wondered if the descriptive words are also needed, or at least would make the paper easier to read to those who might not have a Munsell color chart near at hand. The majority of the scientists on the various boards that discussed the matter and decided that the words were also needed. I went along with and enforced the new rule. A year later, after receiving additional input from more Society members, those same committees and boards reversed that rule. I went along with the reversed rule also.

There have been a few times when I took a side in a discussion. Usually on those occasions I was outvoted. I believe, for instance, when they are both presented in the same reference list, Part 1 of a paper should be listed before Part 2 even in those rare cases where such a listing would cause the two papers to be out of alphabetical order of their authors' names. The decision: alphabetical order prevails over numerical order.

I believe "organic matter concentration" ought to have a hyphen, in the same way as "particle-size distribution." I was outvoted on that, too.

Of course, I abide by the will of the majority in all such cases (even when I disagree). But if you also don't happen to like those rules, don't blame me for them.

The point I'm getting at is straightforward; the things we do in Madison [or most other publishers do in their offices] in the way of enforcing editorial style decisions are not whims of me or any other single person. A number of scientists, writing specialists, and others have gotten together (and continue periodically to get together) in nomenclature committees to determine ways to communicate that will be clear and unambiguous to anyone who is likely to be reading the material.

If you disagree with some of the rules these people have drawn up, you may be comforted to learn that I disagree with a few of them also. But I go along with the rules (in part because I have no choice, but more importantly) in the interests of the clearest possible communication. Writing expert Dick Darby put it this way, "The writer who chooses to submit copy to commercial or professional publications—rather than to write novels or poetry—is ipso facto bound by the rules prevailing in that realm. It is unfair to ask the editor to suspend them."

So, get angry at whatever rules you don't like, but please follow them until you can lobby to get them changed officially. And don't blame me for rules you do not like unless you find out that I actively promoted and endorsed them. And even if I did, I certainly was not the only person, for my opinion alone never prevails; nor should it.

* * *

So, then, should rules never be broken? Not necessarily, as explained in a December 1995, column:

Knowing when to break the rules

Michaelangelo faced a dilemma. He wanted to carve a statue of a large man lying in the lap of a dainty woman, but there was no way he could develop an even remotely pleasing pose if he followed the conventional rules of scale. So he broke the rules; he carved the man in a smaller scale than the woman. That statue, the Pietá, is one of his most famous.

The next time you are in Disneyland in Anaheim, CA, stand just below the train station and look down Main Street. It looks incredibly long. Then move to the other end of the street, stand in the village square, and look toward the train station. It looks like a short distance. This is because the buildings near the train station are taller than the buildings near the village square and the street narrows so as to exaggerate the perspective. (Also, the upper stories of all the buildings are shorter than their first floors, to make the buildings look taller than they actually are.)

Why did the Disney artists visually manipulate the length of the street and the height of the buildings? They had a relatively small area available to build the street, but wanted you, as you arrived at the park in the morning, to see a huge fun-filled area awaiting you and your family. An added bonus was that at the end of the day, when you are tired, it looks like you need take only a short walk to get back to the parking lot, so you might take a little extra time and spend a little more money at the gift shops on your way out. (The artists had more room in Orlando, FL, so the similar street in the Magic Kingdom there is constructed normally.)

In the same way these artists manipulated visual rules, writers who know what they are doing can manipulate grammar and style rules (instead of being bound by them).

The first grammar rule?

Grammar probably grew from the notion, "Do nothing to hinder the reader's smooth passage through the material." The first grammar rule may have been set down after the first grammarian read a sentence like, "I want you to carefully, consistently, and quickly remember this fact." The grammarian told the writer, "The infinitive, 'to remember,' is difficult to follow when separated by so many words. Don't split it like that." The "like that" was dropped somewhere along the line and this rule (in English) became, "Don't split infinitives." (Why can I say this? Simply because split infinitives are allowed in several other languages.)

The following two phrases involve the infinitive "to complete." Which version hinders your smooth passage through the material?

- "to hurriedly complete a difficult task"
- "to complete a difficult task hurriedly."

The second is "proper," it doesn't split the infinitive; but can you honestly say it is more readable than the first? At best it's a tie; but I think the first is a trifle easier to follow because nothing separates "complete" from its adverb, "hurriedly."

Or, consider the line from the TV show, "To boldly go where no one has gone before." I cannot think of a more natural-sounding placement for the word "boldly" than where it splits the infinitive, "to go."

All rules must be approached with common sense. Your underlying goal is that your reader correctly interprets your message. If breaking an arbitrary grammar or style rule helps achieve that goal, then break the rule.

In an earlier column I explained the need to occasionally begin a sentence with a word such as "And" or "But;" breaking a rule you may have learned in grammar school. [The third sentence in the first paragraph of this column began with "So," and two paragraphs ago I began one with "Or."] Beginning a sentence with one of these smaller conjunctions serves to clearly link it to the material that came before it without the cumbersome disruption caused by a more ambiguous word like "however." The only sentence I will begin with the word "however" is in one like: "However many tests you run, make certain the equipment is recalibrated each time."

The first sentence in the final paragraph of the next section will break the rule about subject-verb agreement with "someone–their." This is an example of a practice I do not follow myself, but that has developed in recent years as a way to help make the language gender neutral. More people than you might think use this construction in order to avoid the correct, but cumbersome, "someone–his or her."

Style rules

One of the first instructions I give to new editors is, "Always follow the Style Manual, unless you have a good reason not to." The key words are *good reason*; never break any grammar or style rule *on a whim* or *in ignorance*, and usually not without checking with someone (like a professional editor) first. Generally speaking, when I query other editors about the advisability of breaking a rule in a specific situation, the consensus is that I should not break the rule in question. But I will list a few times when I did.

Our style manual asks you to include the name and address of manufacturers of all products used in all papers. As I've written elsewhere, there are times when common sense requires that this rule be bent or occasionally ignored.

Strict adherence to le Système International d'Unités (S.I.) allows only two time units; seconds and years. More rational heads prevailed in

ASA–CSSA–SSSA, whose style manual allowed also minutes, hours, days, and weeks. Months were not originally allowed because the length can vary from 28 to 31 days; more than 10%.

One day I was editing a paper that contained the phrase, "a 14-month fallow period." I said to myself, "There is nothing ambiguous about this time period; it would be stupid to try to convert this to days, weeks, or years." After checking with an S.I. usage consultant, I retained *months*. I also began to lobby for revising the rule so that it now reads: "The unit, month, may be used for periods of 6 mo or greater in text, tables, and figures." The same unit may also be used for shorter periods of time if accompanied by a word such as "about" or "approximately."

The other day I read an obituary written (or perhaps edited) by someone who adhered to their organization's style rule that required a parenthetical numeral after any written number. This rule makes sense for some legal, scientific, and technical writing, but makes no sense for obituaries: "He is survived by his widow, three (3) children, seven (7) grandchildren, and four (4) great-grandchildren."

Spelling rules

I once planned to describe a tract of land as a "superb suberb" because I thought misspelling "suburb" would make it more alliterative with the adjective. It didn't take me long to realize, though, that (i) spelling it properly would have the same alliterative effect, (ii) people would know for certain what the word meant, and (iii) people wouldn't think I didn't know how to spell a simple word like "suburb."

Deliberately misspelling words is hardly ever a good idea. You can sometimes get away with it for humor. In one of the columns in this book I used the word "redundundundancy" in a subhead for obvious reasons. A few other words occasionally lend themselves to such treatment.

If there is a rule of thumb on misspelling, it would be to grossly misspell the word so there is no doubt it was deliberate; for instance, naming a French poodle "Phydeaux," or sitting down to eat a plate of "phude." George Bernard Shaw once misspelled "fish" as "ghoti;" taking the gh from *rough*, the o from *women*, and the ti from *nation*.

I can think of only one time when misspelling a word in a nonhumorous situation worked as planned. A painter wrote the word *kauphy* on a sign he placed in a store window. People walking by would look at it, scratch their heads, enter the store, and tell the owner, "your painter misspelled something in that sign. I think it's supposed to be 'coffee.'" "I know," the store owner would reply with a smile. "And now that you're here, why don't you pick up some to take home with you?"

Many words in English are spelled illogically (English is, I believe, the only language that can have spelling bees). Even some words that seem to be spelled correctly are not. People are familiar with *nitrogen* as a noun and *nitrogenous* as an adjective, but few are aware that *phosphorus* is a noun and *phosphorous* is an adjective; they incorrectly think *phosphorous* is simply an alternative spelling of *phosphorus*. (I thank Mike Buttress, of Modesto, CA, for reminding me of this.)

Verbless sentences

I'm sure you learned in grammar school that a group of words without at least a subject and a verb is a sentence fragment, not a sentence. Perhaps you asked the teacher if "Go now." could be called a sentence. The teacher said yes, because it had an implied subject, "You."

Likely the teacher added that it would be impossible to have a sentence without a verb. Perhaps you also remember that my July 1995 column (Chapter 3 in this book) contained a verbless sentence. Here's what my favorite reference for things of this sort, Fowler's *Modern English Usage*, has to say about them: "Since the verbless sentence is freely employed by some good writers (as well as extravagantly by many less good ones) it must be classed as modern English usage."

Fowler offers six categories where verbless sentences have been used by good writers. Please note that sentences of this sort cannot stand on their own; they are an extension of the sentence that comes before or occasionally after them. I modified Fowler's examples to make them more appropriate to the sort of writing we might do:

1. *Transitional:* (i) A summary comment on what has gone before: "True, no doubt." "So far, so good." "Of course not." (ii) To introduce what is to follow: "The practical conclusion?" "Finally, on one small point."
2. *Afterthought:* "His research is on a par with Barbara McClintock's. Well, almost."
3. *Dramatic climax:* "We shall face difficulties as we always have done. With honor."
4. *Comment, especially if arch, strident, or meant to surprise:* "We resolved the matter by appointing a committee. A neat solution."

"I used to watch her establish a hypothesis. Fascinating. Brilliant."

5. *Pictorial:* [I can't see much use for this in scientific writing.] "It is an entire street full of shops. Complete with side arcades. And a restaurant. And two snack bars. All piled on top of each other. An entire civilization all to itself. Practically a state."
6. *Aggressive:* "The particular dynamism of the research group from which this discovery came sprang from the rambunctious school of science it nurtured. Defying the conventions. Fighting the taboos. Hastening the inevitable change."

Final word

Please do not misinterpret this column to mean that I recommend flippantly breaking every grammar and style rule you ever learned. Quite the contrary, rules normally should *not* be broken. The only time to break a rule is when (i) you know what you are doing and (ii) you have a more compelling reason to break it than to follow it. It is also a good idea to consult with someone who knows more about the particular subject than you (like I did with the S.I. usage consultant before I allowed use of *months* in a paper the first time, and do with other editors in Headquarters all the time). Another opinion can help you decide whether breaking the rule gains more than it costs in any particular situation.

* * *

If English is not your native language, you have one more consideration to keep in mind as you are beginning to write your paper. In fact, you need to consider this thing if *the American version of English* is not your native language. That consideration is addressed in this column that was published in June 1999:

You need to translate more than just words

Most submissions we receive from Asian, European, and Latin American nations have cover letters that contain a sentence along the lines of "I would be honored if you would give consideration to the publishing of this humble contribution in your esteemed journal." Language even as flowery as that causes no problems in a cover letter. If the same language is also used in the paper itself, though, it presents many problems.

The amount (or perhaps more accurately, the lack) of flowery prose in scientific writing is one of the major sources of difficulty for persons for whom English is a second or third (or whatever) language. Persons who are used to using such flowery phrases when they write in their native language feel uncomfortable when they are forced to write in the clear, straightforward, direct-to-the-point language used in scientific English because the latter seems rough and crude to them.

For reviewers and editors, though, such flowery niceties clutter up a paper and get in the way of its basic purpose, which is communication.

Can be harmful

I have never before written on this subject and, to be honest, had never given it much thought (except when I'm editing a paper written in unusually flowery prose) until a dinner conversation I had with a friend and colleague. He happened to mention that his daughter works in the funding office of a major international service organization that receives grant requests from all around the world. The simple, straightforward requests from some nations are generally dealt with quickly by her board. Members of that same board, though, resent having to wade through the flowery language of the requests from some other nations and sometimes reject those requests before they even finish reading them.

My friend said his daughter is trying to explain to her board that the language they think is so pompous and overblown is the normal way of speaking and writing in those nations. She has been urging her board not to downgrade the request simply because of the way it was written, but to make the effort to wade through the flowery prose and judge the request on its merit. She is also planning to travel to some of those nations to give a series of seminars on how to write grant requests in simpler English so those requests will at least receive a fair hearing from her board. (I gave my friend a full set of reprints of this column to pass on to his daughter to help her prepare that seminar.)

After that dinner conversation I couldn't help wondering if any of our reviewers or editors felt a similar resentment of what most scientists would call "overblown language" when they work on a paper that has language as flowery as the cover letter mentioned at the beginning of this column. If the entire papers are written in the same style as those cover letters, they certainly might. That resentment alone might not cause the rejection of a paper, but the amount of work involved in slimming down the prose so it is suitable for publication might. Editorial board members do not have the time to rewrite papers for authors.

I know there is at least some resentment to the flowery nature of some languages. I once sat in the office of a person as he conversed, by phone in Spanish, with someone in South America. After he finished the conversation and hung up, he turned to me and said, "God, that's a flowery language I have to use when I talk to them."

For writers

If English is your second (or more distant) language, you need to shift your cultural focus as well as your words when you write for scientific publication, just like another friend did. She was born and raised in Canada and spoke nothing but French until she was in college. Now she works in the office of the construction-supply company she and her husband own in Madison. When she first started working in that office, and writing letters to customers and suppliers, she used English words set within her familiar French sentence structure. Neither her husband nor the recipients of those letters appreciated her elaborate way of writing things, and she quickly had to learn to simplify her language.

She told me this change was difficult. She feared that anyone who received such a—to her—abrupt and harsh letter would feel insulted by it. It did not take her long to realize, though, that American readers were not at all insulted by a simple, straightforward presentation of information. In fact, that was precisely what they wanted to receive.

If you experience similar fears about those who will read your scientific English, rest assured that the succinct writing that seems abrupt or rude to you is exactly what your readers want.

Editors and reviewers

If you are reviewing or editing a paper containing language that is far too flowery, and you know its author is someone whose native language is not English, try to be considerate. The author is simply writing in the only way he or she has known since childhood. Before you summarily reject the paper, first try to wade through all the fluff and flower and see if there is some good science underneath it all.

If there is, suggest that the author rewrite the paper so as to uncover that science. If the author needs help, editorial board members have a list of Society members who have volunteered to assist authors with this task. The same may be true for other professional societies as well.

* * *

Please permit a crass commercial plug for my editing and consulting firm at this stage. The head of a U.S. Federal Agency contacted me and asked if I could help him convert a publication about to be issued by his agency to, in his words, "American English." I was happy to be of assistance. The fee was more than that charged for the standard presubmission review, but the department head was pleased with the editing and rewriting I performed for him. If you need that, or any similar editing or rewriting services, contact me at Fine-Tuning Your Writing, LLC. The address is given several places in this book.

Chapter 3

Professional writing

There are people like me who make a portion or even all of their living by writing. Then there are people, like most of you, whose professional development and progress depends on their writing. Each of us can logically be called professional writers.

Are you writing at a professional level, though? Or do you rate as a rank amateur? Hopefully, most of us will rank somewhere between those two extremes, while constantly moving toward the top.

One thing that can help in that regard is if you remember that what you are writing is going to be published, and take into account the particular needs of that publisher.

Sure, you can say to yourself, "I don't care about that, I'll write the material the way I always do and let someone else take care of getting it published." That is the precise stance, though that reveals you as an amateur.

If you are interested in beginning to develop your professional expertise, heed this information, which was first published in February 1994:

Tricks of the trade can make you a pro

No doubt you have seen colleagues or students handle themselves so well in a field, laboratory, classroom, or giving a speech that you remark, "That person really knows what he (she) is doing." Editors notice that about authors as well.

The way you write and your attitude toward other people's suggestions and criticism will create a halo (either a good or a bad halo) for yourself in the minds of editors. Savvy authors will try to create a good halo by writing and acting in a way that indicates they are pros who know what they are doing.

Such authors are rewarded for their efforts; editors eagerly jump at the chance to work on their papers, often setting aside other work to get right at it because they know the paper will be well written and the author will be cooperative.

The by-lines of authors with bad halos evoke groans: Editors have been known to put off working on papers by these authors for several days until they are able to clear the long uninterrupted time period they know the paper will require.

Remember the minor rules

One thing savvy writers do is keep aware of minor rules, and write their papers in a way that does not violate them. One such rule in ASA–CSSA–SSSA publications is that sentences cannot begin with an abbreviation. Non-savvy writers ignore this rule and begin a sentence with something like "KCl fertilization . . . ." Editors correct it, of course, but savvy writers know they can fix it themselves in one of three ways:

- Begin with the word *The*: "The KCl fertilization . . . ."
- Change the word order: "Fertilization with KCl . . . ."
- Write it out: "Potassium chloride fertilization . . . ."

Of these three, the first and second involve far less work than the third, and are the ways professional editors use to change things when needed, particularly if the abbreviation is of a longer and more complicated term than KCl.

Another thing savvy authors keep in mind; the text of papers and Notes published in our journals always begin with a large initial capital letter, so they do not begin their papers with a single-letter word, such as "I" or "A," or an extremely short word such as "It" or "An" or "The."

Also, to comply with the needs of the large initial letter, savvy writers do not begin their papers with a word that must be italicized. When a professional editor gets a paper that begins with a short or italicized word, that person must rewrite the first sentence to bring a word to the beginning that can be used with the large initial letter.

If you write your paper like a pro would in the first place, someone further down the line doesn't need to correct your error.

Forget the incorrect rules

Some authors insist that the grammar rules they learned in grade school or high school are the only way to write. They say, for example, that the word "compare" is *always* followed by the word "with" and the word "contrast" is *always* followed by the word "to." That rule is *usually* correct; comparisons between *like* things require use of the word "with." For example: "One set of data compared readily with the other."

But comparisons between *unlike* things require use of the word "to." The following two sentences (the first from a classic poem) both use "to" after the word "compare," and are both correct because their comparisons are to totally unlike things—a person to a day, and laboratory equipment to a dinosaur:

> "Shall I compare thee to a summer's day?"
>
> "Compared to a dinosaur, my lab equipment is not at all old."

You may have noticed that I began the previous paragraph with the word "but," which violates another rule some people learned in grammar or high school. Remember a teacher telling you that a sentence should never begin a sentence with a word like "but" or "and" because it would be incomplete? That teacher was only partially correct; the *thought* is incomplete, but the sentence is not.

Few, if any, sentences present a complete thought, though. Most sentences are combined with several others into a paragraph that presents a *more* complete thought. Those paragraphs, in turn, are combined into a larger unit (paper, chapter, book, etc.) that presents the most complete thought. But even that is not complete, in the deepest sense of the word; certainly not for all time.

Beginning a sentence with a conjunction like "but" or "and" is perfectly legitimate in a few situations (twice in this paper), and far less cumbersome than the word "however," which some people use instead of one of these shorter words.

Allow us to help you

You say you don't have time to learn (or unlearn) these and all the other grammar, style, and formatting rules? No problem; editors don't expect you to learn them all (if you did, we'd be out of a job). We have a problem, though, when an author:

- Writes a sentence that will not fit into our format and refuses to change it because "it is right the way it is," or
- Does not know the correct grammar or style rule, and demands that an editor follow an incorrect rule.

The simplest solution to dilemmas such as these would be for authors to remember that the professional editing staffs employed by the publisher are there to help authors get their papers published, not to get in the way of the paper's publication. Of course most professional editors aren't scientific experts; authors always have the final say about all scientific aspects of their papers. The professional editors are experts at *publishing* scientific papers, though.

Savvy authors heed the words of a character in a "Peanuts" comic strip many years ago, "Editors are sort of human, too, you know." Editors appreciate it when an author treats them as if they were human beings who know what they are doing. Editors treated in such a manner will return the favor by handling that author's papers as gently and quickly as they can.

* * *

Chapter 4 presents more details of this sort that help make the difference between amateur and professional writers. I want to devote the rest of *this* chapter to another concept common to pros and scary as hell to amateurs; that the art of good writing is the art of throwing words away. More than that, it is often the art and the necessity of throwing some of your *best* words away, as I both explained and demonstrated in September 1995.

Discard the good

After your first few revisions there may be nothing really bad in your paper, but there probably is too much of a good thing.

Poor writers find reasons to keep it all: "This paragraph is interesting." "This sentence is particularly well written." "I want readers to know this information, too."

Good writers ignore all such thoughts and ask themselves one question only, "What in here is essential to the subject of this paper?" The essential stays and the merely relevant goes.

This is not simply a matter of discarding the good and keeping the better. In at least some cases the discardable material has been written with more flair, and might even be more interesting than the essential material. If it's not essential to your paper's subject, though, get rid of it.

This month I will show you how to spot nonessential material in your writing and next month I'll show how much of it I throw away as I produce one of these columns.

Without question, information that is not essential to one paper might be essential to another. Such material should be moved to, or even become the basis of, another paper. That is not the sort of material I refer to here, however.

Story that didn't make it

In early drafts of another column I used a delightful story to illustrate a point. As I refined the column, though, I could see the story was interrupting the flow of the rest of the factual material. I liked the story, though, and moved it out of the paper into what writers call a sidebar. After a few more revisions I deleted that sidebar.

That story is still in my files and may someday appear in another column. I doubt it, though; the piece joins many other of my lines that have not made it into print because they did not contribute to the overall subject matter as well as other lines that were retained.

This used to bother me, as I suppose it bothers a beginning surgeon to cut through perfectly good skin to get at a medical problem; or bothers a beginning gardener to rip a perfectly good corn plant out of a bean field.

In surgery, the bottom line is life and health of the patient. In gardening, the bottom line is the crop. In writing, the bottom line is essential information.

How to do this

How can a writer separate the essential from the merely relevant? A good place to begin is with your first sentence—throw it away. If this sounds too drastic, at least consider throwing your first sentence away. When I began writing professionally I discovered one of the easiest ways to improve the opening of a paper was to throw away my first sentence and use the second sentence as my beginning. I assumed my writing was unusual until I began editing other people's work and found the same thing in at least half of those papers.

I don't know why, but most writers' first few words are often vague and unfocused. Perhaps it is a little like clearing one's throat before speaking. By the second sentence, or certainly the second paragraph, the passageways are clear and good material is flowing from brain to paper or word processor.

Look at anything you have written recently (scientific paper, departmental report, or personal letter) and see if your second sentence makes a better opening statement than your first sentence. I'll give 50–50 odds it does.

After 30 years of writing professionally my batting average is better than 0.500, but I still throw away 10 to 25% of my first sentences. (This column was unusual in that I deleted its *first three paragraphs*. I present them at the end of this paper so you can see for yourself what I mean by deletable material.)

As well as improving the opening of your paper, discarding the first line is a good mental exercise. It is a way of emphasizing to yourself how serious you are about publishing only essential material. After you've thrown away your first line—and found your paper better as a result—it becomes less painful to spot and throw away other nonessential material.

Redundundundancy

Did you say the same thing more than once? Simple repetition is essential in a speech, but usually not a good idea in a written piece. Where you find yourself repeating the same words, change them so you present the same idea several different ways. Then readers who do not understand one version might understand another.

This entire article, in fact, is little more than a host of variations on its first sentence: "After your first few revisions there may be nothing really bad in your paper, but there probably is too much of a good thing."

Be ruthless

I recommended in my November 1993 column [published in Chapter 4 of this book] that you be your own worst critic. In this column I'm suggesting that you be your own most ruthless editor as well.

Go through your paper like a physician performing exploratory surgery or a gardener looking for weeds. When you find a sentence that contributes to the flow of ideas, keep it and enhance it if you can. When you find a relevant-but-nonessential phrase, sentence, or paragraph, ask yourself if it is contributing to the life of your paper. If not, get rid of it even if it is interesting, written with more flair, or in any other way superior to the essential material. No matter how pretty it might look or sound, it is sucking life out your paper by distracting the reader from the material you are trying to convey.

The original first three paragraphs of the previous column

Your first sentence, or sometimes your first paragraph, often is not as good a beginning for a paper as the material you write after it. My first draft of the previous column began with a two-paragraph story. As I refined the column I decided the story was too long, and shortened it to a single paragraph. Still later I deleted it because I realized that it detracted too much from the other material. In short, it was relevant but not essential. Those two false beginnings are reproduced here. Please note the continuing refinements in the final sentence of these attempts that became the first sentence of the published paper.

First version:

I've never had difficulty choosing between good and bad; the dilemma comes when I need to choose between two good things. About 30 years ago, when I lived in Dubuque, IA, I was offered two different jobs. One was in Madison, WI, the other back at Iowa State where I went to school. Both positions had about equal pay and prestige, and I considered both cities nice places to live. I chose the newspaper job in Madison and never again thought about the job I did not take or the city of Ames. A year-and-a-half later I left that job in Madison for a distinctly better one, working for ASA–CSSA–SSSA.

After we had lived in Madison for a while my wife and I decided the time had come to move our growing family from an apartment to a house. We found two houses that fit our needs. Both were attractive, well built, and in good neighborhoods. We chose one and forgot the other.

Writers face dilemmas like this all the time. After the first few revisions there is nothing really bad in a paper anymore, yet there is too much of a good thing.

Second version:

I seldom need to choose between good and bad; but I often must choose between two good things. In life situations, such choices have

been between one job or another, one city to live in or another, one school to send our children to or another, and one house or car to purchase or another.

Writers face dilemmas like this, too. After the first few revisions, there is nothing really bad in a paper, yet there is too much of a good thing.

---

* * *

If the examples just given were not enough for you, the next column, published in October 1995, one month after the previous one, tells how much I throw away when I write a column (or just about anything else).

Writing, the fine art of throwing words away

The science fiction writer Douglas Adams, when asked to describe his writing process, said the main thing he does is delete words.

For example, shortly after he began to write the radio script based on his *Hitchhiker's Guide to the Galaxy* books, someone asked how he was doing. "Off to a good start. I have 25 pages completed." A few days later, in response to a similar question from someone else, he replied, "I have 18 good pages." A few weeks later he submitted his first 12 pages to the program's producer. This, of course, is the same technique I described in last month's column; discarding the good.

This month I'll tell you how much of one of these columns is typically discarded before you see it. As soon as possible after I get an idea for a column I enter it into my computer. My first entry generally contains 300 to 500 words. This column began with 338 words.

The next time I work on an embryonic column I revisit those thoughts and flesh them out with information I've look up from various sources, pertinent examples, quotations from other people, etc. By the time I've finished with this session the typical column contains 800 to 1,500 words, and this column contained 747 words.

Still later I'll expand a typical column to perhaps 2,300 words. This one was shorter, it reached only 1,465 words at that stage.

By this time the column contains essentially all the information I want to present, and my task switches from creating the column to getting it ready for publication. I begin by throwing away bad phrases and combining or deleting redundant sentences. By the end of that session a typical column will be about 1,800 words long. Still later I'll begin the process I described last month as separating the essential from the merely relevant, and reduce length to between 1,400 and 1,600 words. This column had 1,221 words at that point.

Then I circulate it to the other managing editors in the building and my boss for their critique and input. They occasionally suggest additional information, which can bring the paper back up to about 2,000 words. When their additions are substantial, they may remind me of something I need to add myself, and the paper may grow to as many as 2,400 words. This column reached 1,255 words at that point.

Then I begin my final refinement, and the number of words declines to 2,000, 1,800, then 1,600 words. By the time I'm satisfied with a column—a work of art (or science) is never finished, merely abandoned—it usually contains 1,200 to 1,400 words; occasionally as low as 800, sometimes as large as 1,600. This one contained 1,220 words when I handed it to *Agronomy News*, whose editor often shortens it more.

You may wonder, given what you have read thus far, how I can write a column during the four weeks between issues of *Agronomy News*. I easily could if this was the only thing I did, but these columns are not even a part of my position description; I write them in my spare time (a few minutes per session) after hours and between other duties. I try to initiate at least one new column each month and have five or six in development at any one time. Some take as few as three weeks to write, others require six months or more.

About the first of each month I select one column from the two or three closest to completion and concentrate on getting it ready for publication. By the 8th or 10th of the month I'm ready to send it around the office for review, hoping to get it back by the 15th. By the end of the month I give it to the editor of *Agronomy News*. I began writing this one on 18 May and handed it to *Agronomy News* on 23 August.

One becomes two, or three

If I cannot get a column shorter than 1,600 words, an unofficial limit I have set for myself, I have two choices. The simplest is to ask the editor of *Agronomy News* if he has space for a longer-than-usual column in the upcoming issue. He normally does.

If space is limited, though, I'll search my files for a column I can complete before deadline, and save my longer column for an issue with more available space.

My other choice, especially when a column contains 2,000 or more words, is to see if I can divide it. If so, I'll split off a publishable unit of 1,200 or so words, give it to the editor of *Agronomy News*, and plan to get the rest ready to publish in the next issue.

The three-part column earlier this year on considering the needs of the reader [Chapter 1 in this book] began as a single column for the March issue. At the time I split off the first part, I thought one additional column would be all I needed, and concluded the first part with: "I'll expand on the concept of writing for the reader next month and give you some specific questions that can help you write papers or speeches for a variety of people in a variety of situations."

As I worked on that second part, though, more and more ideas came to the fore and I could see the column would need to be divided at least one more time. By the time I had the second part ready to publish, in fact, I was concerned whether I could squeeze all the remaining information into only one additional column, or if I would also need a fourth. I practiced what I preach, though, and was able to throw away enough material to allow me to complete the discussion in the May issue.

The paper you are reading now was planned from the beginning as the second segment of a two-part series.

Why work so hard on a paper?

An individual's investment in writing can vary widely. My reward for these columns is purely psychic. I get no extra pay for them and would have to immediately stop writing them if so ordered by my employer. I write them because you have indicated in cards, letters, phone calls, and personal contacts that you need and appreciate them. I have never produced anything during my more than 30 years of professional writing that was as well received as these columns.

A typical scientist, on the other hand, spends hundreds of thousands of dollars on the research project described in the paper, then spends an additional several hundred to more than a thousand dollars in page and reprint charges to publish it. Future research grants and sometimes promotion or tenure decisions may be riding on the paper. Thus, you and I both have good reasons for investing the time and energy it takes to make our writing as fine as we possibly can.

While there are a few writers who get it perfect the first time and never have to do a bit of rewriting, most writers agonize about every word, throw most words away, and retain only the essential. Use whatever technique works for you, but if you think your writing needs improvement, consider trying the technique Douglas Adams and I use.

* * *

As you can see, one sentence in that column is no longer correct. I wrote in 1995 that I would immediately stop writing the column if ordered to do so by my employer. Four years later, when that actually happened, I decided instead to take an early retirement so I could continue the column as a newsletter, and offer it as one of the services of the consulting firm I was establishing.

* * *

Sometimes the line between being a professional writer and causing more harm than help is thin. The professional editing staff at ASA–CSSA–SSSA, after dealing with a number of problems from a number of authors, asked me to write the following column, which was published in September 1993. I am inserting it exactly as it appeared. Some of the information regarding electronic processing is out of date, but if I were to update it to the situation today, it would be out of date soon again anyway, due to the speed with which things are changing in that area. My best advice to you regarding electronic submission is to check the current situation with your publisher before you submit.

Your help sometimes doesn't help us here

Most authors try hard to cooperate with Headquarters editors. Such cooperation is not only appreciated, it also is mutually beneficial. It helps us to do our jobs and it helps your paper to get into the journal as quickly as possible.

There are some things authors do, though, thinking they are helping us, that actually hinder and slow our work. I decided to mention them this month in hopes of improving the situation.

Reference section abbreviations

A major problem we face is trying to fathom the abbreviations authors make up for publication names in their Reference lists. Please, if you do not know the library abbreviation for a publication, give us its full title. (The abbreviation for American is Am., not Amer.; the abbreviation for pollution is pollut., not poll.) Both of these are fairly easy to decipher, but many others we receive are not.

It can take us considerable time to page through title after title in our reference sources to see which publication the author might have had in mind. Often we have to ask the author to supply the name of the publication and the name and address of the publisher. When we receive the full name, we usually can find the proper abbreviation in our references.

If you give us the full title from the beginning, though, we can easily find the publication, then delete the unnecessary words or letters from your text to create the proper abbreviation.

Don't send a replacement manuscript

When you get questions from a Headquarters editor about a reference or a request to clarify some other point in your manuscript, please answer those questions directly. Since word processors became common, some authors will incorporate the changes into their manuscript, run off a new version, and send it to us to replace the old. No doubt these authors think they are doing us a favor by giving us a nice clean copy to work from.

Actually, they are causing us substantial extra work. You see, we already edited your manuscript before we asked those questions. If, instead of answers to our questions, you send us a whole new manuscript, we must do one of two things—both equally time consuming.

- Set aside the version we have edited and edit your new version of the manuscript.
- Put your new version alongside our edited version and search line-by-line for the answers to our questions so we can transfer them to the edited version. We must also look to see if you have made any other changes (and there is a good chance that some of those changes will be overlooked). Those changes we find, we must also transfer to the edited version of your manuscript.

In other words, when you send a new manuscript instead of simply answering editorial queries, you force us to edit your manuscript twice. By contrast, when you send specific answers to our specific questions, all we need to do is to pull the edited version of your manuscript from the file, plug in your answers, and your manuscript is ready for the typesetter in a matter of minutes.

Doublespacing means doublespacing

All authors know we ask for doublespaced, typed manuscripts, and virtually all authors send in manuscripts with their texts prepared that way (except for those that seem to think 1.5-line spacing is just as good—it isn't). For some reason, though—maybe because they are printed in smaller type in the journals—we often get Reference sections, table captions, and occasionally Abstracts that are typed singlespaced.

We realize it takes more paper to type everything doublespaced, but we need all material doublespaced. You see, we do more than just read manuscripts here; we make spelling or grammatical corrections, we insert or take out periods or commas, we write instructions to typesetters about spacing and type faces, we write notes to ourselves about things that require further checking.

In short, when we edit a paper, there are any number of things we might need to add to or take out of a reference section, table caption, or Abstract. (Reference sections, in fact, often need far more of this sort of work than any other part of a paper.) It is nearly impossible for us to do any of this legibly when the typed lines are jammed tightly together.

Editorial boards, under special circumstances, can make an occasional exception to the "doublespace everything" rule. Authors should never violate the rule without checking with an editorial board member or Headquarters first.

When to send a disk

We can have a number of problems when you send us an electronic version of your manuscript on a computer disk. The most obvious problem is that sometimes the disk arrives so damaged that it cannot be used. This is one of the reasons why we require a hard copy version with each manuscript submitted on disk.

There are other reasons for the paper back-up as well; tables and math formulae often do not translate well from one word-processing software to another. (Virtually all tables and math must be hand-keyboarded by the typesetter.) Also, there are some things—such as whether a paper has too many or too few subheadings—that are more visible when a person can see the whole manuscript at once on paper than when it is viewed only a few lines at a time on a screen.

The second obvious problem with disk submission is that some of you send the wrong kind of disk. Headquarters operates on a PC-based system. At least for now, we have no way to convert and use disks from Macintosh machines. Save yourself (and us) some postage—don't even bother to send us Mack disks. The only thing we can do with a Mack disk is return it to you.

Occasionally the disk version we receive does not exactly match the paper version of the manuscript. Then we need to determine which version was approved by the editorial board for publication so that we do not end up publishing the wrong one.

A far too common problem with disk submissions is that the disk arrives after we have edited the manuscript on paper. If we knew a disk was coming, we would have waited for it to arrive. (Rather than make us wait, it would be better that the disk arrives with the manuscript.) When a disk arrives after we have already edited a manuscript, we probably will not edit the manuscript again on a computer. The slight gain in accuracy of typesetting would not be enough to justify the expense of double editorial work.

Still another way you can cause problems with disks is when you send us a whole new disk in response to editorial queries. This is similar to what was discussed above with hard copy revisions. You force us to either set aside the work we have already done and edit the manuscript again or take time to sort through the new disk to glean the bits of information we need to know to add to the manuscript we have edited. Depending on the software (some will allow the computer itself to compare and highlight differences between two disks), this can range from a difficult to an extremely difficult task. But either way, it is a problem we would rather not have to face at all.

* * *

By the end of 1994 I had published quite a few original columns, and we had seen a significant improvement in the quality of submissions to ASA–CSSA–SSSA journals. We also saw a few annoying problems that had not improved. So I wrote this sort of an update, which was published in January 1995. I hope you will forgive the level of annoyance evident in this entry, it was totally unlike most of my writing.

The bugs that used to bug us are still bugging us

If I were asked to summarize in one sentence the 35 columns I've written to date, I would borrow an old sentence from someone whose name I cannot remember, "The three C's of writing are clear, concise, correct."

This month, I want to revisit some problems I've mentioned in earlier columns, partly for those who may have not seen them, partly because they are important enough to be repeated, and mainly because I continue to see these problems appearing in papers.

In one column I wrote that anytime you repeat a word (we sampled a sample) there is probably a better way to present the sentence. Consider the following sentence that came in a paper written after that column was published: "Selection intensities were selected to give an overall selection intensity of approximately 25%." The author could just as well have written, "Overall selection intensities were about 25%." (Remember also my column that suggests replacing longer words, like *approximately*, with shorter words, like *about* whenever possible?)

In another column I mentioned that any construction involving the word *respectively* should be carefully considered because it is often difficult to follow? Here's a sentence from a recent Fellows biography that demonstrates this: "He completed the M.S. and Ph.D. degrees in cytogenetics and plant breeding in 1965 and 1967, respectively." Does this mean that he received an M.S. in cytogenetics in 1965 and a Ph.D. in plant breeding in 1967; or were both of his degrees in a combined curriculum of cytogenetics and plant breeding?

Tell me; What is wrong with the following sentence? "Forty grams of moist soil in 100 mL plastic tubes (o.d. = 31 mm), adding 60 mL of distilled water." If you said it has no verb, you are correct. Someday I will be a guest on an Oprah or Geraldo Show when the subject is *Sentences that lack something, and the writers who love them.*

Then there are the long sentences. I swear that some authors have never heard of a period, exclamation mark, or question mark; all they know how to use are commas, colons, and semicolons. Their sentences go on and on and on and . . . . I have broken some of these sentences up into two, three, and sometimes even four or five sentences. If there is a long distance between the subject and the verb in a long sentence, it is just that much more difficult to read. I'm not a sadist, so I won't subject you to examples of any of these.

Then there are the authors who are probably otherwise nice people; loved, no doubt by their dog, children, and spouse; who insist on not doublespacing their manuscripts, or who doublespace only part of it. Doublespacing the text but not the reference section is not enough. Doublespacing the table and not its caption is not enough. If you have ever attempted to edit a single-spaced author-paper documentation, or a single-spaced reference section, you will readily see the necessity of doublespacing. One-and-a-half line spacing isn't enough; it has to be double throughout.

I just mentioned the author-paper documentation section. Have you ever heard of it? It is the small paragraph that comes between the author by-line and the abstract of the paper that tells the name(s) and address(es) of the paper's author(s), any required documentation, such as experiment station number, an asterisk (*) followed by the words "Corresponding author" and the word "Received," with a blank space after it so we can fill in the date we receive it. They are not that difficult to write, believe me; just look at the beginning of any paper in any of our journals and you will see an example. Then follow that example and write something like it at the beginning of every paper you write. They have, after all, been required since the issuing of the 1988 edition of our style manual.

In case you hadn't heard, about 15 years ago, all ASA, CSSA, and SSSA journals except *Journal of Production Agriculture,* (and it didn't exist then) became S.I. publications. That means that we use le Système International d'Unités of measurement. We allow certain specified exceptions, and the author always has the right to incorporate another unit of measurement parenthetically after the S.I. number. But you must use S.I. If you don't, we will either try to convert it for you here at the office (and may well get it wrong) or send part or all of the manuscript back to you for correcting. Make it easy on yourself (and a great many other people who will be reading and working on your paper) and use S.I. the first time around.

And I don't care what style you have always used when you create a table or a figure or what style some other publication or publisher uses; if you are writing a paper for ASA, CSSA, or SSSA, please use the table and figure style promulgated by those three scientific societies. If you don't the first time you write the paper, you can bet that the paper will be in that style before it is published. Again, save yourself and a great many other nice people a great deal of work and get it right the first time.

For at least 4 years, the three societies have frowned on too much use of author-made-up abbreviations; especially those that can be ambiguous (equidistant plant spacing and embryo protein synthesis would both be EPS). Try not to use any abbreviations (except those specifically allowed by the Style Manual), but if you must use some legitimate abbreviations, please define them in a separate, unnumbered footnote at the beginning of the paper, not parenthetically within the text.

I've noticed that some people still seem to like to overuse the word *over*. As a special gift, I conclude this column with a short story I wrote just for them. I tried to use *over* as often and in as many ways (correct and incorrect) as I possibly could. I hope all who like the word *over* enjoy it.

(And I hope the rest of you see what an ambiguous word it is and from now on use a word that is more specific to the particular situation.)

Over 18 people, jammed themselves into a plane (which was over its capacity) and flew over a field because they were invited by over 16 relatives to come over to their homes and have a barbeque. While the over 16 relatives were waiting the arrival of the overloaded plane that was carrying the over 18 people (which was overdue) they put hamburgers over a grill that was placed over hot charcoal to begin preparing the meal. The over 16 people, of course, needed to make sure that the fire over that grill was not overly hot, so the hamburgers would not be overly done or overly dry and taste overly tough. The pilot of the overloaded plane applied the brakes and reversed the engines as soon as his wheels came down over the runway, but because the plane was overloaded, it caused an overheating and failure of the braking system. The plane overshot the runway, and smashed into over five homes just over the fence over the end of the runway. Over 25 people were killed in the terrible tragedy and the resulting fire burned for over a day, even though there were over 15 pieces of fire-fighting apparatus and over 30 firefighters attempting to overcome the damage caused by the blaze. Over 10 of the people at the scene were overcome with smoke and had to be taken over to the hospital. Over four ambulances had to make over three trips apiece to get all the injured and those who were overcome with smoke over to the hospital. And with that sad news, this story is over.

* * *

With that out of my system, I returned to my once again pleasant and slightly whimsical self, as evidenced by this paper, which was published in July of the same year, 1995:

How's that again?
Did I miss something?

Occasionally as I read scientific or other literature I am awestruck by a beautifully written sentence. Other times I get a mental workout and then a chuckle from an incomprehensible sentence. I first think, "How's that again? I must have missed something." After reading it several more times, though, I realize I didn't miss something, the author did.

Sometimes the problem is as simple as a missing verb. "Forty grams of moist soil in 100 mL plastic tubes (o.d. = 31 mm), adding 60 mL of distilled water."

[There is such a thing as a verbless sentence. It sounds like a contradiction, and such sentences make no sense out of context, but good writers use them occasionally—poor writers use them frequently. I placed a properly used verbless sentence in this paper as a demonstration. See if you can find it. If you can't, I'll tell you at the end where it is.]

Incomplete comparisons

Advertisers often fail to complete comparisons. When they write: "20% stronger!" or "50% smoother!" they want you to think, "Wow, what a great product!" They do not want you to ask, "Stronger than what, a piece of cotton thread? Smoother than what, a rocky seacoast?"

Scientists occasionally fail to complete comparisons, too: "Each of the three designs described is a variant of a split-plot design, but these variants are all different." Does the author mean different from each other, or from the way they should have been set up?

"[Then we exposed the test soil] to different tillage systems." Different from the way farmers till here on earth? Probably not; perhaps a number is missing: " . . . 12 [or whatever] different tillage systems." Then again, *time* may have been the missing element: ". . . tillage at three different times of the year."

The authors may have avoided specifics because the combination of numbers and times (or other variations) was too complex to delineate in a single sentence. If that was the case, they could have used, " . . . a variety of tillage times and systems." This is both understandable in itself and also tells the reader more details will follow.

Run-on sentences

I had to read a 73-word sentence several times before I finally realized it was not *one*, but three sentences that had been awkwardly hooked together with "in which" and "and in which." Deleting those connectors (and making two other minor changes) revealed the three sentences. Here it is, as written, with strikeouts and capital letters inserted to show how I divided one sentence into three: "The objective of this study was to determine the response in seed yield from a tandem selection scheme. ~~in which~~ The first step was to delete lines that were undesirable for morphological and developmental traits from each of the two single-cross populations by using independent culling. ~~and in which~~ The second portion of [the] tandem ~~selection~~ was to select for greater plant N content at the R5 stage among the previously selected lines."

These sentences could be further refined, but they are understandable after these minor changes.

Illogical construction

The problem with the next example is not so much the choice of words as it is poor word arrangement or perhaps logic. A reader can't tell whether the authors' focus is salt-tolerant plants, new crop development, or nonproductive lands: "A search for plant species with economic potential as a crop that is already somewhat adapted to saline conditions is a prerequisite for the proper utilization of these nonproductive lands."

I think the subject is nonproductive lands and the authors meant something like this: "These saline soils will not become productive until we find plant species capable of growing in them, and then modify those plants so they produce an economically viable crop."

To prove this was not a problem of word choice, I'll restate my interpretation in the authors' own words: "A prerequisite for the proper utilization of these nonproductive lands involves a search for a plant species that is already somewhat adapted to saline conditions and that has economic potential as a crop."

Ideas, not just words

Anyone who knows how to spell (even imperfectly) can put words onto paper. The purpose of writing, though, is to transfer a writer's thoughts into the mind of a reader. Good writing is difficult because writers have only one tool to convey their thoughts; symbols, which include words, numbers, punctuation, and illustrations. A good writer can accomplish a great deal with these few symbols. The job of the writer is to select, arrange, and try to package these symbols into a form readers can comprehend.

How can authors know for certain that their specific arrangement of symbols will successfully transfer their thoughts into their readers' heads? In a word, proofreading. But by that word I mean far more than a quick check with your co-authors to make sure they agree with the technical details, and one last quick skimming of the paper yourself before you send it off to the journal. The kind of proofreading I refer to here involves two distinct steps.

Read your paper aloud

How can you know if readers will interpret your words the way (or ways) you want them interpreted? Read them aloud. Reading aloud forces you out of the writer mode and into the reader mode. Readers expect to find information in certain places. You will be amazed at the number of awkward constructions and poor word choices you didn't even see when you read silently that you immediately catch when you read aloud.

I used the parenthetical "(or ways)" in the previous paragraph because authors sometimes (rarely in scientific writing) use words that have more than one meaning. Three paragraphs ago I wrote: "A good writer can accomplish a great deal with these few symbols." I chose the phrase *a great deal* because it: (i) *means* (in this context) to transfer as much information as possible from the mind of the author into the mind of the reader, and (ii) *connotes* (in this and other contexts) production of something both valuable and memorable.

Poets have mastered the use of words with more than one meaning or connotation and the best professional writers can usually make them work. Everyone else should normally stay clear of them. First of all, you need to know all the meanings and connotations of a word and consciously want to convey all of them.

Second, if you accidentally misuse a word, people may forget what you meant, but they will never forget your mistake. Forty years ago, in a seventh grade health class, I watched a film about vaccinations. At one point in the film the narrator said something like, "Johnny got a little prick." For the next five minutes the boys and girls in the room whispered and snickered and paid no attention to the film. That gaffe is the only part of the film I remember. Would you want your paper remembered for something like that?

I faced a different sort of word-meaning problem with the original version of a sentence in the sixth paragraph of this paper. I had written: "[The test soil] was exposed to different tillage systems." This sentence contained no obvious errors, but for some reason my brain misinterpreted "the test soil" as "the soil test" almost every time I read it. I didn't want readers to make the same mistake, so I changed my bracketed part to: "[Then we exposed the test soil] to different tillage systems." I don't know why, but I never misread the phrase after changing it from the subject to the object of the sentence.

Ask someone else to read it

Once you have read your paper aloud, caught and corrected all ambiguities, and are certain your entire paper is the best you could possibly produce, you are ready for the second proofreading step; ask another scientist to read it. Don't ask a co-author; select someone who has never seen the paper and, ideally, is only somewhat familiar with your work. (This way you are selecting someone typical of those who will read your *published* paper.)

Most authors neglect this step, submit their paper to a journal, and hope the review-editing process will accomplish the same result. It may, but these authors have relinquished the advantage of an informal, close-to-home reading.

Heeding the advice of a disinterested but knowledgeable person before submission will enable you to submit the best possible paper. At worst, this will do no harm. More likely, it will simplify and speed publication. Also, consider the value of the halo effect. Journal editors likely will be impressed by and grateful for a high-quality paper and give it their kindest possible treatment. Without this second proofreading step, those same editors might resent the additional time and work it takes to decipher a not-so-perfect paper.

Regardless of who gives the first outside opinion of your paper (close-to-home colleague or journal reviewer and editor), do not be surprised or offended by a request that you decipher or clarify a sentence or paragraph you thought was perfect. This happens with every writer. Believe me, I know; *every* writer.

When readers ask for clarification, they are telling you that the symbols you used did not transfer your ideas into their heads. Authors refuse to heed such advice only at their own peril.

[Did you find the verbless sentence? It was, "In a word, proofreading." under the subhead, "Ideas, not just words." I may, in some future column, explain more about the use of this type of sentence.] Actually, I did write that explanation. You can find it in Chapter 2 (p. 44–45).

* * *

The next column was prompted by a request from several editorial boards, who expressed concern that many authors (and more than a few editorial board members) did not understand the difference between a Note and a full research paper. It was published in February 1995.

What is a Note? How is it different from a paper?

Beginning in 1988, editorial groups within ASA–CSSA–SSSA agreed that Notes were to be used to describe only one of three things:

- A new apparatus
- New research techniques
- Observations of unique (usually unrepeatable) phenomena.

Notes are also supposed to be "normally less than two printed pages." These rules are violated more and more, with Note lengths now averaging between 3 and 6 published pages. At their various meetings in Seattle (1994), the editorial boards were asked to decide whether to enforce the "normally less than two pages rule" or let notes get as long as the author wanted. The boards voted to keep a closer eye on length, and will encourage authors to cut back on the length of papers that seriously exceed this limit or to consider converting them to full-fledged papers.

When faced with a similar question a few years ago (when the length of Notes was just beginning to get out of hand), the editors of *Agronomy Journal* decided they would normally publish longer pieces as full papers, rather than as Notes. The *Soil Science Society of America Journal*, on the other hand, said it would publish the information as a Note, regardless of its length.

You may have noticed other variation among the journals as well; the *Journal of Environmental Quality* calls them "Short Communications." Most of our journals publish notes in their separate section in the back of the journal; *SSSA Journal*, though, publishes them in the back of the particular section of the journal where they are most appropriate. Thus, that journal has Division S-1 Notes, Division S-2 Notes, and so forth.

These variations are minor. The main similarity of Notes is their subject matter. Regardless of what they are called or in which section of a journal they appear, notes are supposed to be short and to deal only with (i) new apparatus, (ii) new research techniques, or (iii) observations of unique, usually unrepeatable, phenomenon.

What is a new apparatus?

Many researchers appreciate learning of new and better equipment they can use in their work. Perhaps the two watchwords editorial boards use in deciding whether to accept a paper in this category is (i) whether the piece of apparatus is new and valuable to scientists, and (ii) whether the paper is an actual scientific paper or a thinly disguised advertisement.

What is a new research technique?

Similar to information about new products, information about new ways of doing things can be valuable. Without information of this sort, researchers could spend a great deal of time reinventing the wheel.

Information of this sort is so valuable that this category has been divided. Information on computer software is published in its own section in *Agronomy Journal* (for scientific and research software) and the *Journal of Production Agriculture* (for production-oriented software), and may soon be published in other journals as well.

Information on computer models is now published in its own section in *Agronomy Journal*. Our other journals will publish papers on computer models as well.

What is a unique observation?

This is perhaps the most controversial criterion for Notes. There has been a great deal of discussion as to what constitutes something that is unique and what is something that merely ought to be studied more; if not by the person who discovered it, then by someone else who has the time or inclination to do the studying.

I hope the dilemma involving research topic ideas has been at least partially clarified by the new procedure announced in this column in the March 1994 issue of *Agronomy News* [directly below in this book]. Scientists who notice something interesting they do not have the time or inclination to pursue further can submit a one-doublespaced-typewritten-page account of it to the most appropriate of our journals where it will be evaluated for publishing as a "Research topic idea letter to the editor."

* * *

And here is that column from March 1994:

A place now to submit research topic ideas

From time to time scientists will see something *they* don't want to pursue, but that someone else might. Although this information could be passed on to a colleague at a departmental meeting, phoned to a friend in another state, or mentioned in passing at a professional meeting, many scientists have wished for a more formal way to get the information out to a wider audience. Two recent examples:

- A scientist applied various chemicals to tall fescue in an attempt to elicit a phytoalexin response. One of the chemicals, iodo-acetate, increased the amount of chlorophyll by at least 100-fold. The scientist was interested in disease resistence, not basic plant physiology, and did not pursue the matter.
- In a set of soybean plots near a wooded area, deer one night ate all but one of the soybean lines down to the ground.

The scientists who observed these happenings thought about submitting them to one of our journals as Notes, based on the third criteria for Notes: "Observations of unique (usually unrepeatable, such as hail or frost damage) phenomena." The first was not submitted because the author believed it would be turned down; the second was submitted and turned down. The reasons:

- The information on chlorophyll could be repeated and therefore was not unique. Thus, it should be repeated prior to publication, even though the scientist who discovered it was unable to pursue the work.
- The soybean that appeared to be deer resistant may or may not be eaten by deer if it were the only strain of soybean in a field and the deer had no choice. Premature release in the literature of information on resistance might cause more confusion and controversy than enlightenment.

Because of complaints from these and other authors, some journal editorial boards at their annual meetings in Cincinnati in 1993 discussed the definition of "unique phenomena."

"It is something that happens during an eclipse," said J.L. Steiner, agroclimatologist at the USDA-ARS, Bushland, TX, and an associate editor of *Agronomy Journal.* Her notion that "unique and unrepeatable" means precisely that was echoed by most members of most boards. The conclusion: Possible research subjects are not suitable topics for Notes.

Members of the Editorial Policy Coordinating Committee agreed with this stance, but also agreed with the authors that *some* vehicle is needed for publishing research topic ideas. They asked me to investigate the matter and find a way to publish them.

Others publish similar information

One often reads in letters-to-the-editor sections of chemical journals about unexpected and unexplained explosions when chemicals were mixed together in a new way. The chemists who writes these letters generally do not know why the chemicals exploded, but write the letter to warn other chemists working with those chemicals to be careful until more is known.

The letters-to-the-editor sections of many medical journals include clinical observations. Physician–readers of these journals take these observations for what they are worth and some elect to investigate the matter further.

I wondered if the letters-to-the-editor section of our journals (perhaps redefined and expanded) be a place to publish the sort of information described at the beginning of this paper. Between meetings in Cincinnati I asked that question of Jerry Hatfield, editor of *Agronomy Journal.* He said the idea had merit and urged me to pursue it further.

No reason why we can't

Current editorial policies allow this use of letters. The *Publications Handbook and Style Manual* says the following:

> "All the societies' journals publish letters to the editor. Letters may contain comments on articles appearing in the journals *or general discussions about agronomic research*, and are limited to one printed page." [Emphasis added.]

A few people might consider research topic ideas more of a specific than a general discussion, but this doesn't seem to be stretching the rule so far as to render it useless.

One printed page is probably too long for information of this sort; a quarter of a printed page (one doublespaced typewritten page) would be a more suitable limit for research topic ideas. (Most probably would need no more than three or four sentences.) Perhaps they should be given a subtitle, like "Research Topic Letters," within the letters to the editor section to call better attention to them. That would be up to the editor of each journal to decide for that journal.

I see little danger of abuse of our journals or the letter-to-the-editor system by publishing letters of this type. The remainder of the paragraph from our style manual, quoted above, states, in part:

> "Letters must be approved by the editor of the journal and may be subject to peer review."

All research topic letters would be given at least a cursory technical review by the journal's editor (and possibly others) so that readers' time and journal space would not be wasted with trivial material. For those concerned about publications of this sort inflating a person's vitae, letters to the editor generally are not given a great deal of credit on a vitae.

In short, I can see many advantages and few, if any, disadvantages if journals began to use letters-to-the-editor for research topic ideas.

I sent an early draft of this column and a mail ballot to the editors-in-chief of our three societies and the editors of all the journals for their comment and possible approval. They liked the idea; agreed that it was a change in practice, not of policy; and approved that change. The new practice, therefore, can begin immediately.

All of our journals will, of course, continue to accept the traditional letters to the editor. Members of the editorial boards of those journals have, in fact, talked about ways to encourage more letters. Perhaps this new category for letters will help in that regard.

Thus, from now on:

- If you have a comment about an article appearing in a journal or a general discussion about agronomic science, crop science, or soil science research, submit it as a traditional letter to the editor of the appropriate journal.
- If you want to report an observation "of a unique (usually unrepeat-able, such as hail and frost damage) phenomena," (as well, of course, as a new research technique or new apparatus) submit it to the appropriate journal as a Note.
- If you find something in your work that merits further research that you don't want to or are unable to perform, write an account of it in no more than one doublespaced typewritten page and submit it as a Research Topic Letter to the editor of the appropriate journal.

* * *

The final column to be reprinted in this chapter deals with a pet peeve of mine; that of people who know nothing about science spouting off about science and scientists not offering any comments about either those blatherings or failing to present the truth of the matter because they do not believe it is a proper thing for a scientist to do.

I received a lot of criticism from my bosses when I sent this manuscript out for peer review within the office, and I was forced to send it to an outside-the-office review and add an editor's note; both of which actually strengthened the paper:

> Editor's note: A prepublication copy of this column was circulated to the Editors-in-Chief of the three Societies, the current Presidents of ASA and SSSA, the President-Elect of CSSA, and a former president of CSSA who is now the Executive Vice President of ASA–CSSA–SSSA. All seven unanimously approved publication of this column, and several made suggestions that strengthened it beyond the original version.

Call for speculation in scientific writing

Until a couple decades ago scientists would often conclude their papers with a few words of speculation on the next logical step of the research or the possible long-term meanings of their discoveries. More recently many scientists frowned on speculation, and you could find it only rarely in our journals. Now more than a few scientists again see a need for more widespread speculation.

One told me, "Wouldn't it be nice if we could still publish the sort of papers they used to publish in *Crop Science*? I learned a lot from them when I was in graduate school, and got a lot of good ideas for my own research projects."

What sort of speculation?

If it wasn't for speculation there would be no research; what is a hypothesis but a speculation? This is not the speculation I refer to in this paper, though. What I mean is to make some speculation at the end of a paper about deeper meanings to the research results.

I thumbed through the 1950 volume of *Agronomy Journal* looking for examples in papers written by people who are acknowledged today as respected scientists. I selected three papers based only on the name of the scientist, and found speculation in all three:

> G.F. Sprague and B. Brimhall, "Relative effectiveness of

two systems of selection for oil content in the corn kernel," Agron. J. 42:83–88. This paper describes work with the then-new breeding technique, recurrent selection, on corn. In the final three paragraphs the authors speculate on ways to modify the recurrent selection technique to screen for other traits in corn and how to modify the technique further so it could be used for other cross-pollinated crops.

R.G. Blaser and N.C. Brady, "Nutrient competition in plant association," Agron. J. 42:128–135. This paper investigates why the feed quality and composition of pastures decline with time and use. The paper concludes that some plants are helped and some are hurt by fertility management practices. The authors then speculate: "The results obtained herein also suggest that the physiological approach in species improvement should be pursued. It is possible that plant breeders should make an endeavor to isolate strains of legumes that are potentially good nutrient absorbers. . . . Nutrient competition among species may also be of help in explaining the degeneration of pastures and meadows from desirable to undesirable species."

L.F. Graber, "A century of alfalfa culture in America," Agron. J. 42:525–533. After the author of this historical paper describes the previous hundred years of alfalfa culture, he concludes: "With large stocks of foundation seed for rapid increases of new and superior varieties for northern and southern latitudes, and for humid and arid climates as well, it is difficult to envision the magnitude of the impacts that alfalfa will have on the agriculture of North America in the years ahead. Suffice it to say that when all these achievements are applied to make alfalfa better suited to the environments of the new frontiers of its culture and when such environments are adjusted to the needs of alfalfa in the eastern half of the United States and Canada, alfalfa is destined to become the greatest leguminous forage for hay, silage, and pasture in North America."

If scientists as highly respected as George Sprague, Nyle Brady, Roy Blaser, and Larry Graber were speculating in their papers in 1950, there must not have been anything considered wrong with speculation. (I don't know who B. Brimhall was. He or she was identified as a research assistant on the paper and there is no one currently on the membership rolls with that name.) These were not the only authors who speculated. The Watson and Crick paper that introduced DNA to the world concluded with a remarkably prescient speculation as to the ultimate significance of their discovery.

Not any more, but why?

Why did this sort of speculation, so common three decades ago, fall by the wayside? For one thing, scientists' practical need to hang onto their thoughts so they can research them instead of giving someone else the opportunity first. This concern is certainly valid, as evidenced by the quote in the second paragraph of this paper.

A second reason is that some persons, particularly those who review papers, seemed to believe that speculation beyond demonstrated findings is not proper. To use the words of Sgt. Joe Friday of the old *Dragnet* TV show, these reviewers want, "just the facts, Ma'am."

There was never any ASA–CSSA–SSSA policy against speculation. The policy, as stated on page 24 of our *Publications Handbook and Style Manual*, reads:

> "Speculation is encouraged, but should be reasonable; firmly grounded in observation and subject to tests. It must also be identified apart from the discussion and conclusion. Where results differ from previous results for unexplained reasons, possible explanations should not be labored. Controversial issues should be discussed clearly and fairly."

The first three words of that statement are certainly clear enough. One could argue that the words that follow could be interpreted as either a license or a straightjacket. Many peer reviewers and probably also editorial board members in recent years have interpreted the rule more like a straightjacket than a license. Many authors vigorously push the idea of speculation for their papers, then take the opposite stance when they review someone else's paper. In short, although our Style Manual at least allows, if not encourages speculation, our editorial process has all but forbidden it.

Speculation has not stopped

Speculation about the results of scientific research has not stopped. In fact, it is commonplace today. The difference is that instead of being done by scientists in peer-reviewed journals it is done by politicians, network news anchors and reporters, radio talk show hosts, and TV and movie celebrities. These people have ready access to the media, can talk convincingly, and have little or no credible opposition, so a scientifically illiterate public believes them.

The general public "knows" that today's scientific agriculture is not nearly as good for the soil, crops, or consumers as the type of agriculture practiced 100 years ago. The general public also "knows" that anything developed by a scientist or made or refined in a factory could not possibly be natural, and therefore must be bad. At least a part of the public also "knows" that anything with chemicals in it is bad. I once asked the chemical makeup of a fertilizer and was told, "Oh, it doesn't have chemicals, it's all natural."

The people doing the speculating today have done more than simply manufacture irrational fear. Consider the Alar scare of a few years ago or the number of other products that have been banned with little or no evidence that they are dangerous and little or no consideration of benefits vs. risks.

Such beliefs as astrology and fortune telling are treated as authentic and scientists who try to present evidence otherwise are ignored. As an example of how low the esteem of scientists have fallen, just consider popular entertainment. Have you ever seen a TV show or movie where a scientist was depicted as anything other than either so mad as to want to blow up or take over the world or so flaky as to do something stupid like shrink or grossly expand the size of his own children?

Should scientists speculate again?

One could argue that because some people are speculating irresponsibly, then scientists ought to get back to speculating responsibly in their papers. I would be inclined to push for that myself, but it doesn't make for a terribly logical argument.

One could hope that if scientists start speculating responsibly in their journals that somehow this truth would seep out and bring common sense to the rest of the world. One could hope that, but whether it would actually happen is problematical.

The only rational argument for scientific speculation is that it is and always has been a logical and rational part of science. Just because I happen to believe that doesn't necessarily mean it should happen, though, so I asked some experts what they thought. I created a questionnaire and circulated it to the Editors of all our journals and the Editors-in-Chief of our three Societies. The questions and a summary of their answers accompany this paper.

Before I get to their specific answers, I want to say that they enthusiastically supported the idea of again allowing thoughtful scientific speculation in our journals. When it took me longer than planned to complete this column, many of those Editors and Editors-in-Chief prodded me because they wanted the information out. I took advantage of the delay, though, to discuss the subject in St. Louis (1995) with Technical Editors and the full editorial boards of the journals I am associated with, and also received strong encouragement from them for the concept.

Survey answers

Every Editor and Editor-in-Chief agreed speculation in scientific papers is occasionally appropriate, depending on the nature of the paper, and most said speculation is always appropriate and should be encouraged. (They could check more than one answer to each question.)

When asked who was most qualified to speculate, all said the scientist who did the work. Most indicated that other scientists familiar with the work would also be qualified to publish speculation about it.

How much speculation should be allowed? "Be much less rigid; allow speculation when appropriate," received the most votes, closely followed by, "Not only allow speculation, but be so bold as to actively seek it from authors who have not supplied it in the paper, especially if the paper's finding appears to be a breakthrough."

Your Editors and Editors-in-Chief unanimously agreed that speculation was more likely the sign of a leading journal in its field than of an undisciplined journal that has little respect.

Will speculation harm a journal, its sponsors, or anyone? The unanimous response; "Nobody would be hurt and many will be helped, because speculation is part of the scientific process. Thoughtful speculation may well open whole new avenues for research."

There were three replies to; "The proper place for speculation is:"

- "At the end of a regular scientific paper." 20% agreed.
- "In a separate review or interpretive paper in a scientific journal." 20% agreed.
- "Both scientific and interpretive papers, but to varying degrees." 60% agreed.

If speculation were to be encouraged, who would police it and define the limits? Two-thirds of the votes were for, “Technical Editors, Associate Editors, etc. as a standard part of the review process.” The other third of the votes went to “Peer Reviewers.”

Conclusion

Scientific journals have historically been an appropriate place for thoughtful, scholarly speculation, and there appears to be no reason why the practice should not be allowed to resume in ASA–CSSA–SSSA journals. Quite the contrary, many persons see many reasons why speculation *should* resume.

Science is suffering because the only speculation being done about it is by people who know little or nothing about science. Scientists should get back into this discussion and contribute their expertise wherever and whenever appropriate. In the words of Dr. H.H. Cheng, SSSA President, “We should encourage speculation to stimulate thoughts, but not to sensationalize the consequences. Perhaps this is what differentiates scientific speculation from irresponsible speculation.”

The return of credible scientific speculation to the discussion will not immediately silence irresponsible speculation. It may give the public another side to the question, and, who knows, maybe ultimately the truth will win out. Even if it makes no difference at all in the public debate, that is no reason why it shouldn’t be done.

Should speculation be required in all papers? Of course not. First of all, not all papers lend themselves to speculation. Also, some authors may not feel comfortable speculating.

I hope, though, that if some authors believe it would be appropriate to include some clearly marked scholarly speculation at the end of their next paper that they do it. And I hope that the peer reviewers and editors who help authors prepare that paper for publishing will not summarily reject the speculation, but rather see if it meets the criteria outlined in the style manual and, if so, will seriously consider leaving it in.

Those criteria require that any speculation be reasonable, firmly grounded in observation, subject to tests, clearly identified as speculation, and if it discusses any controversial issues, does so clearly and fairly.

[The survey and its results begin below.]

These are the multiple-choice questions asked of ASA-CSSA-SSSA

Editors and Editors-in-Chief in the survey discussed in this column. Respondents could check more than one answer to each question. The answers in boldface type received the highest number of responses. See the article for their complete answers.

1. Speculation in a scientific paper is:
   a. Never appropriate
   b. **Occasionally appropriate, depending on the nature of the paper.**
   c. **Always appropriate and should be encouraged.**
2. The only people qualified to speculate on the findings of a paper are:
   a. Nobody.
   b. Historians 50 years after the paper is published.
   c. **The scientist(s) who did the work.**
   d. **Other scientist who are familiar with the work.**
   e. Journalists who are familiar with the work.
   f. Jeremy Rifkin. [This received one response, "Him, too."]
   g. Politicians, attorneys, religious, so-called environmental activists, and others with big professional mouths who know nothing about either the work, the scientific process, or science in general.
3. ASA–CSSA–SSSA journals should:
   a. Never allow speculation in any form under any circumstance.
   b. Warily allow a minute amount of speculation, but only by scientists who are at least 65 years old and who have published no fewer than 300 peer-reviewed nonspeculative papers in the subject area.
   c. **Be much less rigid; allow speculation when appropriate.**
   d. **Not only allow speculation, but be so bold as to actively seek it from authors who have not supplied it in the paper, especially if the paper's finding appears to be a breakthrough.**
   e. Let everyone and anyone speculate on any subject regardless of their background or expertise in the field.
4. Speculation is the sign of:
   a. **A leading journal in its field.**
   b. An undisciplined journal that has little respect in its field.
5. The people who will be hurt by speculation are:
   a. Sponsors of the journal, who will lose prestige if the journal

would even stoop to doing something so unscientific as to speculate on anything. If a statement has not already been proven beyond a shadow of a doubt, it is by definition not scientific and therefore not appropriate for inclusion in the pages of a journal.

b. Sponsors of a journal, who will lose prestige if the speculation proves to be incorrect.

c. Sponsors of a journal, after they get sued because someone took action based on the speculation and it didn't work.

d. Other journals, who are not as forward-thinking as the journal that allows speculation.

e. **Nobody, and many will be helped; speculation is a part of the scientific process. Thoughtful speculation may well open whole new avenues for research.**

6. The proper place for speculation is:

   a. Newspaper letters-to-the-editor sections; never in scientific journals.

   b. **At the end of a regular scientific paper.**

   c. **In a separate review or interpretive paper in a scientific journal.**

   d. **Both scientific and interpretive papers, but to varying degrees.**

   e. None of the above.

7. If speculation were to be encouraged, who would police it and define the limits?

   a. Ad-hoc committees consisting of the most knowledgeable people in any given subject area, appointed by the society president on a case-by-case basis.

   b. Journal editors.

   c. **Peer reviewers.**

   d. **Technical editors, associate editors, etc. as a standard part of the review process.**

Chapter 4

Truly great writing

In case you haven't noticed by now, this book takes the opposite order of most "how to write" books. In the first four chapters I present the ideal, to give you a goal to strive toward. Only after that ideal is planted firmly in your heads, and the goal firmly established, will I begin to deal with the sort of things you see in most writing books; things such as grammar, style, reference lists, and that sort of thing.

This book will also contain a great deal of information that not only is not traditional to the "how to write" books, this book will contain quite a bit of information you've likely never seen anyplace (unless you subscribe to my monthly newsletter). One of those is the "Don't disguise your writing" segment you will find in this chapter.

Before I get to that, though, I begin this chapter with a statement of philosophy first published in April 1999, and updated for this book:

How to learn how to write

I do not believe it is possible for one person to teach another person how to write. One person can teach another *about* writing, but the only way to learn how to write is to write.

Don't get me wrong, learning *about* writing is good, in fact, it forms the base for learning how to write. The things you learn *about* writing are

---

a little like the forms a contractor sets up. Great writing is the concrete you pour into those forms. Only you, though, can mix and pour that concrete that is your writing and only you can create the conditions that allow that concrete to harden. Without those forms, your concrete will harden into an unidentifiable worthless lump instead of a magnificent structure.

You and I need to know these rules of grammar, composition, style, and so forth, plus many other things, if we are ever going to write well. So most every month I try to read something myself on how to write well. I subscribe to a fairly expensive newsletter titled *Editorial Eye*, and a more popularized magazine titled *Writer's Digest*, and I own probably as many books on writing as any other professional has on his or her profession. But as good as all that sort information is, it will not *teach* me how to write well (or you, either), anymore than concrete forms will ever become a permanent structure.

Write

If you want to become a great writer, the *only* way you can make that happen is to write, and then write some more, and then write some more. I read recently that nobody can be considered a really good writer until he or she has written at least a million words. As you might suspect, as soon as I read that I looked back across my writing career and tried to tally up how many words I have written; about 4 million to date. (The original version of this column alone has contributed about 93,000 words to that total—93 columns times an average of 1000 words per column. Then add in close to 200,000 for a nearly completed novel and another 6,000 words for each issue of my newsletter—15 issues at the time these words were written. In fact, I have probably passed the 4-million-word mark.)

But am I a really good writer? When I look at *some* of the things I have written I think so. When I look at some other things I almost want to laugh (or cringe) at a piece's awkwardness or pomposity or opacity. In short, I'm much better than I was when I started in this business 40 years ago, but I'm not as good as I hope I eventually will be. So I continue to write, partly to earn a living, partly to inform others, and partly for the selfish and pleasurable purpose of sharpening my own skills.

I encourage you to do likewise. Keep writing, and from time to time take an honest look at something you have written. If it was good, admit it, congratulate yourself, and try to do as well or better the next time. If it was bad, try to figure out why it was bad and learn from that.

Read

The second most important thing you can do to is to read what others have written, and I don't mean only "how-to-write" books and my newsletter. The most important thing a writer can read is the same sort of things that he or she writes—scientists should read scientific papers and books, romance novelists should read romance novels, etc.

That ought not be a problem for you because you are already reading scientific papers and books to keep up with your profession. You likely also read a newspaper or two and a few magazines. Most of you try to find some time to read fiction and some poetry now and then, too. All of this is good, especially if you make use of it as a teaching tool.

How do you do that? Do you ever, after reading something, say to yourself, "I can write better than that"? If so, good; that shows you not only are learning the differences between good and bad writing, you also are learning where you currently fit on that continuum.

Do you, after reading some other thing, say or think, "Man, I wish I could write like that"? Good too, particularly if you study the piece and try to figure out what it was about its writing that you liked. Maybe it was the author's clever use of words. Maybe it was the smooth way that one sentence flowed into another. Maybe it was the use of analogies that made clear something that was otherwise opaque. Maybe it was simply the clarity of the prose; after a single reading, you completely understood what the author meant.

I hope after the first couple times you make discoveries such as these you will say to yourself, "I'm going to have to try that in the next thing I write." As you learn more and more of these things you will reach the point that you begin to use them automatically because your subconscious mind is assimilating the information and making it a part of you.

Learn

Look for good writing wherever you can find it. Much earlier in my writing career I often listened to a man named Jim Mader on local (Madison, WI) radio and TV. He was a disk jockey who also made a lot of radio and TV commercials. He had a way of presenting information that was at the same time gentle and incredibly memorable.

For example, he told in one of his commercials for an automobile dealership the story of a woman buying a car who told the salesman, "My son is an auto mechanic and he recommends that I buy a Honda."

"Why is that," the salesman asked. "Does he work on them?"

"No," she said. "He works on lots of other brands, but he's *never* had to work on a Honda. He says that must mean they make really good cars."

For the first 10 or so years of my professional writing career I tried to write the way Jim Mader talked. (That's not as unusual as you might think. On Stevie Wonder's first records he sounded a lot like Ray Charles. On the Beatles' first records they sounded a lot like Buddy Holly.)

I don't consciously copy Jim Mader anymore, but even now, many years after Jim's death, I can still see a little bit of him in my writing. I also see bits and pieces of others, too; teachers I've known, colleagues I've worked with, friends I've loved, and (of course) authors I've read.

I have stated in previous columns (Chapter 6, p. 167 in this book) that the purpose of writing is not to put words onto paper or into a word processor; rather, it is to take an idea from the writer's mind and to insert it, as intact as possible, into the reader's mind.

While this is a definition of *good* writing, I have since come to realize that truly great writing goes far beyond this. Truly great writing takes a portion of the writer himself or herself and somehow makes that a part of the reader. What is a writer made of? Among other things, a lifetime of experiences, education, and knowledge that somehow add up to what is usually called wisdom.

Make it you

Do you want to become a good writer? Write a lot. Do you want to become a great writer? Read not just to learn information, but also to learn how that information was presented. Do you want to become a truly outstanding writer? Become as completely aware as you possibly can of everything within and without you and draw from the best of it every time you write. Thus, the only true answer to the question, "How long did it take you to write this?" is your age at the time you wrote it.

The interesting paradox to all of this is that the more of yourself you put into your writing, the less of yourself people will see. That story of Jim Mader's I quoted above did not, on the surface, tell you a thing about him, it sold cars. You have to read it much more deeply before you come to realize that only someone with a lifetime of knowledge of what it takes to sell an idea could have put together something so memorable.

With those thoughts in mind, here is the way a typical reader might react to poor, good, and outstanding scientific writing.

Poor: What was it, again, that he did?

Good: Gee, he's smart. He set up that research project perfectly.

Outstanding: That's a fascinating piece of research; I think I'll try to duplicate it.

* * *

So what, then, is the difference between good writing, and a level of writing that can truly be called excellent or outstanding? Is there any one trait that makes all the difference? Surprisingly, the answer to that question is a firm "No," as I outlined a month later. From the May 1999 column, and updated for this book:

What makes great writing great?

It may not look like it at the outset, but if you stick with it, you will see that this column is a follow-up to last month's on how to become a great writer.

In a cabinet in ASA–CSSA–SSSA Headquarters was (it is now in my own office) a picture taken in 1986 at an unnamed horse-racing track. In that photo a number of ASA–CSSA–SSSA members and their spouses are gathered around the owners and jockey of the winning horse. The owners, a man and a woman, clearly stand out from the rest of the people in the photo. Every now and then during first few years that picture hung in that cabinet I tried to figure out what makes those horse owners—two obviously wealthy people—stand out so prominently from the rest of the people in the picture.

First I thought it might be the woman's hairdo, but then I realized her hair is no more attractive than that of a couple other women in the picture. Then I thought it might be the clothing worn by the couple, which is certainly nice, but no nicer than the clothing of one or two of the other men and women in the picture. It isn't their slimness, a couple other people in the photo are just as slim. The couple in the center certainly stood in a self-assured manner, and I thought for a while it might be that, until I realized that one or two other people around them stood with as much poise.

Try as hard as I might, I just could not put my finger on any one trait or item that made those two people stand out so clearly from all the rest of the people in the photo. I finally realized I was looking at the picture the wrong way. Every person in the photo, except the two in the middle, possessed one trait that was striking or attractive or both. Only the man and woman in the middle possessed *all* of those traits. They didn't just have good-looking hair or good-looking clothes or a regal carriage and poise or whatever; they had good-looking hair *and* good-looking clothes *and* attractive carriage *and* were poised *and* attractive *and* so forth.

How this applies to writing

The difference between truly outstanding writing and all the rest is something like the people in that picture: Any piece of writing that has one outstanding thing about it cannot be called outstanding writing.

Likely you have noticed yourself that one scientific paper you read might describe the setup of the experiment well, but be weak on reporting the findings; another paper might report the findings well, but be weak on the implications of those findings; one paper might have perfectly presented prose, but have little of substance to say; and another paper might report something of extreme significance, but report that information in a way that is difficult to read. In another, the spelling may be perfect, but you cannot understand the sentences. In still another, the text may be understandable, but the tables and figures are not.

To be good, a piece of writing must have at least two or three good things about it, and to be outstanding, essentially everything about it must be done right.

The outstanding scientific paper is one that presents a topic of some substance in a way that is clear and memorable. The perfect scientific paper, if such a thing is ever published, will present a significant subject logically, with clear prose, and enlightening tables and illustrations.

If you want to compare your most recent (or next) paper to that elusive goal of perfection, judge it not by how far it misses that goal, but instead by how close it comes to it.

The outstanding paper

If you want your writing to be considered as outstanding, remember the lesson of last month's column; the more you write, the more you will learn about writing, and—if you work at it—the better you will get at writing.

At the same time you are improving your writing, you are likely also advancing in your profession, which means you are drawing nearer to the cutting edge of research in your field and increasing the odds that you will be the person who makes that truly significant discovery. So each paper, if you work at it, will take you another step closer to that goal of perfection. If you have been improving your ability to write at the same rate that you have been pushing back the envelope of knowledge in your field, you are moving ever closer to that goal of publishing something that is outstandingly significant.

Moving target

Neither you nor I nor anyone else will never achieve perfection, of course, because it is a moving target. And presumably, much of the moving of that target will be done by people like you as you advance the knowledge in your particular field of research and publishing.

No need to be discouraged, that is the way it usually is with the leading figures in all fields. Wilt Chamberlain was the greatest basketball player of his day, but not as good as Kareem Abdul-Jabbar, who came along a couple years later. And, of course, Michael Jordan did things on a basketball floor that Kareem never dreamed of. But someday, someone else will not only equal, but eclipse everything Jordan has done.

The identical performance that won someone an Olympic gold medal 12 years ago probably would not get the person who repeated it today even a bronze medal. The same performance that won a perfect 10 a few years ago might make only a 9.3 this year.

Frank Lloyd Wright, the greatest architect of the 20th Century (at least), was often asked which of his buildings he considered to be the greatest he had ever designed. Even shortly before his death at age 92 he always gave the same answer: "My next one."

Anyone with an attitude like that will get far closer to that elusive goal of perfection than will the rest of the people who are content with "good enough."

* * *

Not surprisingly, a number of persons wrote me about this column when it was first published. But instead of commenting on the philosophy of writing, they commented on the photo they had not even seen:

"Of course the people in the center stood out, all the people on either side were looking at them." No, everyone was looking at the camera and saying "cheese."

"Of course the couple stood out, the jockey and the horse was between them." No again, there was no horse in the picture and the jockey was between that couple and another couple beside them.

If those who wrote me had been as critical of their own writing as they were of mine they might have been able to improve their own writing. I explained how in this next column, from November 1993:

Be your own worst critic and improve your writing

Every writing teacher agrees that if a person will let a piece of writing sit for several days, then take another look at it, some obvious errors and misstatements will immediately come to the fore.

The elapsed time has helped you to distance yourself from your writing. During the days between the time you have written the piece and the time you read it, other events will have happened in your life to distract you from the piece of writing. Also, the piece of writing will have faded some in your conscious memory. With it not so fresh in your mind, you can approach it as more of a stranger. (Also, as mentioned in Chapter 2, it allows time for the subconscious part of your brain to work on it.)

While this system is good as far as it goes, there are some mental tricks you can use to help you improve this distancing and enable you look at your piece even more objectively, with more of a critical eye. Each writer has his or her own bag tricks; I will list some of the most common ones here. You won't want to or cannot use all of them, because what works for one person may not work for another. Pick and choose from these, and possibly come up with some of your own so that you develop a system that works best for you.

Use your imagination

One mental trick that works well, if you have a good imagination, is to pretend that someone else, instead of you, is reading your piece. You can choose whomever you want to do the reading, but it helps if it is someone with a slightly offbeat or obnoxious personality. Suppose, for instance, you picked the comedian, Jay Leno:

[*Audience still chuckling from previous joke.*] "You know, ladies and gentlemen, you can learn some *amazing* things by reading scientific journals. [*Snicker from audience.*] Yea, really. Why just the other day I learned that; here, let me read it to you. 'High-protein plants had a higher level of protein than low-protein plants.' [*Several loud chuckles.*] Now isn't that amazing! Can you imagine, that somebody is actually getting paid to make a startling discovery like that!?" [*Louder laughing and some applause.*]

After that it won't take you long to change that sentence to "High-protein plants averaged 59% more protein than low-protein plants."

If you have a difficult time imagining someone else reading your piece, read it yourself, but *imagine it was written* by someone else; preferably someone you don't like. Before you begin to read, say aloud or to yourself: "Let's see how bad a job that dummy did this time." That's what is meant by the expression, "Being your own worst critic."

Look for specific things

I'm a tennis fan and beginning player. Usually when I watch a professional tennis match on TV, I do so purely for the enjoyment. But sometimes I try to learn something. For instance, for two or three games or sets I might watch only the way one player moves his or her feet. Then, for a while, I watch the way one player stands as the other prepares to serve. Then I'll watch one other specific aspect of the game.

You can do something like that when you critique your own writing. Read it many times, concentrating on one thing each time. If it is a paper about soybean, for instance, go through it once and look only for the word soybean, so you can make certain you always spelled it *soybean* and not *soybeans*.

Another time, check the ending of each sentence to make certain you used the correct punctuation mark. Were there questions that ended with a period or statements that ended with a question mark?

Still another time, make certain that every sentence had a subject and verb in it, that they agree, and that they were placed as closely together as reasonably possible.

Take as many trips as you need to through your paper, checking out only one thing each time, until you have exhausted the possibilities.

As a critical reviewer

So far, it may seem like I have dealt primarily with the writing style and grammatical points. That is not the case; all the above mental tricks will help you deal with the scientific questions as well. Use any or all of those tricks to be your own scientific reviewer as you anticipate the questions peer reviewers will ask, and answer them ahead of time.

- What did you set out to discover? Is that presented clearly and completely enough in the paper that other scientists can understand it and appreciate its significance?
- How did you set up and conduct your research? Is it presented

in enough detail that a competent scientist would be able to duplicate your work?
- Was the work itself designed well enough that you were able to determine what it was you were looking for without other traits confounding your results?
- What were those results? Were they presented clearly and completely?
- Did you find what you expected to find or were there some pleasant (or unpleasant) surprises? Are these presented in such a way that the reader will have no difficulty finding, understanding, and appreciating them?
- What are some of the longer-term implications of your findings? Are they presented as clearly as possible?

Conclusion

You might say, "All of that takes a great deal of time. Why can't I just send it off when I am finished writing?"

That is a good question, and I can think of two replies. The first is flippant; you are not finished writing until you have done all of the above.

The second, more practical, answer is that all of the above needs to be done before the paper can be published. If you spend an extra week on your manuscript before you submit it, you could save yourself and reviewers and editors countless weeks during the subsequent review and editing steps.

I conclude with two questions for you to answer in your own mind. Wouldn't it be great to one day get a note from a scientific editor, in perhaps as little time as three or four weeks after you submitted a paper (the current ASA–CSSA–SSSA record is 18 days), that reads, "Acceptable as submitted, no revisions needed."? Then, a few weeks later, wouldn't it be great to get a note from a professional editor in Madison (or wherever the publisher of the journal happens to have its office) that reads, "Thanks for an excellent manuscript. We have no editorial queries and will not need to make any editing changes."?

* * *

One thing not mentioned above to look for as you critique your writing is to see whether what you have written looks like and reads like what it actually says. The reason why this information was not included is because I didn't discover it until about 4 years later. I wasn't even sure what to call the problem after I discovered it. The name I finally came up with was *Disguised Writing*, which I defined and explained in this column, written in October 1997 and updated for this book:

Don't disguise your writing

Here is the beginning of a sentence from one of our publications: "Research has shown that although K has relatively little influence on improving stand establishment, yield and stand survival . . . ."

What does this much of the sentence look like to you? To me it looked like a list of things that potassium has relatively little influence on, and I expected the next word in the sentence would be *it*. But the next word threw me, because it was *are*. Immediately I knew I needed to shift mental gears because the sentence with that word in it could not be a list.

I call this type of error *disguised writing*, which is not the best term because it almost connotes something the author did deliberately. But I cannot think of a better term for a sentence that looks like it means one thing when it actually means something else.

The sentence's following three words confirmed my judgement that I was not reading a list, and gave me a second dilemma, because those words were, *highly dependent on*. My mind's task suddenly had become one of figuring out a structure that would logically connect *highly dependent on* and *relatively little influence on* in the same sentence.

Here is the full sentence as it was published: "Research has shown that although K has relatively little influence on improving stand establishment, yield and stand survival are highly dependent on an adequate K supply." As you can see, the mental shifting of gears does not take long, and after a person reads it a time or two the sentence can be interpreted in a way that makes sense.

Stumbling blocks

In an earlier column (published in Chapter 6 of this book) I mentioned the need for an author to do the best possible job of getting his or her ideas onto paper, and the concomitant job of a reader to do the best possible job of interpreting those words so as to understand the author's ideas. Some have called this coding and decoding, which is not a bad choice of words.

I focused that earlier column on reader responsibility. This time I want to focus on the responsibility of the author.

The author of the sentence under discussion did a bad job of coding his or her idea, choosing a code normally interpreted by a reader as a list of items. The writing is not disguised so badly that the reader cannot figure it out, but the decoding of this sentence is more work and takes more time than necessary.

Disguised writing errors are far more common than you might realize. The disruption is seldom consciously noticed because it usually takes the reader only a fraction of a second to shift mental gears and decode what the author meant. Although the individual errors may not be noticed, they (or the lack of them) are the reason why some papers are a joy to read and some other papers are not. Each little impediment, each little bump in the road, makes the entire road just that much rougher to traverse.

Removing those blocks

The simplest way (but not the best, as I will show) to correct the confusion in the example sentence would be to move the center phrase to the beginning: "Although K has relatively little influence on improving stand establishment, research has shown that yield and stand survival are highly dependent on an adequate K supply."

No confusion here, the sentence reads logically and does not look like a list. The only problem is that this simple change also changes the meaning of the sentence. Only the second assertion in this revision is supported by the phrase *research has shown*. Since both assertions in the author's version are so supported, they probably should be supported in any correction, too, unless that correction is made by the author, who alone knows the true meaning of the sentence.

The only way I can think of to logically support both assertions would be to recast the sentence as two sentences: "Research has shown that K has relatively little influence on improving stand establishment. One should still apply it, though, because that same research has shown yield and stand survival are highly dependent on an adequate K supply."

This version is longer (37 vs. 27 words) but would be read more quickly than the original because these sentences need be read only once to be understood.

One more example

Rather than burden you, I will give just one more example: "Would that everyone could write as well as the best of us." This construction is fairly common, especially in speech, because the speaker is able to give the proper inflection so the listener can comprehend the meaning. When it is written, though, the sentence starts out looking like a question, such as, "Would that person with the loud radio please turn it down?"

The third word presents a disruption to the reader, though. I cannot think of any logical question in which *everyone* would follow *Would that.* So at the third word the reader abruptly discovers he or she is not reading a question, and needs to regroup and make a second effort at decoding the author's writing.

To correct the problem, a person could rewrite this sentence two ways. One would be to go ahead and make it a (rhetorical) question, "Wouldn't it be nice if everyone could write as well as the best of us?" The other would be to remove the appearance of a question, "If only everyone could write as well as the best of us." A variation of that might be, "I wish everyone could write as well as the best of us."

Look for disguises in your writing

As you are making your final few trips through your paper before you submit it, consider each sentence carefully. (I find that reading the text aloud helps me spot this type of error.) If any sentence causes you to stumble, you can bet that anyone less familiar with your work than you will stumble even more than you did.

Determine exactly why the sentence caused you to stumble. If the problem was one of disguised writing, use the techniques I described in the two examples in this column to correct your prose.

* * *

Another thing to be on the lookout for as you review your writing is the sort of errors that began appearing after people began writing their papers on computers instead of typewriters, as outlined in this paper from August 1997:

Word processors introduce errors

Word processors allow an author to go back into a page again and again to correct, refine, polish, improve, add to, delete from, or whatever else might be needed.

As such, word processors have made me a better writer than I could have been before they were invented. Back when I used a typewriter, if I saw anything that was less than perfect on a page I needed to weigh the advantages of correcting it against the possible disadvantage of introducing an unrelated typing error somewhere else on that page.

Once in a great while I could type a replacement word, phrase, or sentence on a separate piece of paper, cut out a small strip from the paper containing those words, paste that piece on top of the word or words I did not like, and photocopy the page. Those corrections worked only when they were essentially the same length as the original. Substantial deletions were close to impossible to make without completely retyping the page. And, of course, any change that would lengthen a section would mean retyping several pages, so usually was not done.

Not all rosy

Although word processors allow such changes without introducing typographical errors in another part of a page, they can allow you to introduce a whole new type of error where you made the change.

If you notice and correct all these errors before you submit your papers, I congratulate you; you are a better proofreader than many of your colleagues. Your professional editing staff at Headquarters often finds these sorts of errors in manuscripts. If you haven't noticed errors of this sort in your writing, you might recheck something you've written recently. Maybe you either are not proofreading as carefully as you should or are not using your word processor to its full potential.

Nature of the errors

These errors happen when a person makes any of the revisions mentioned above, and fails to delete all the words (or all the parts of words) the revision is meant to replace. The result is aberrant letters, words, even phrases or whole sentences; all of which are extremely difficult to detect. Spell checkers will usually not catch them, neither will grammar checkers.

Here's an example of the sort of error I mean, from the title of a self-study course I was writing. My first title read, "Faster-than-normal corn planting sometimes hurts yields, but not always." That was too cumbersome, so I intended to shorten it to "Faster-than-normal corn planting doesn't always hurt yields." I correctly inserted all the needed new words and deleted all the unnecessary old words, but neglected to change one word I did not delete. So instead of what I wanted, the title then read, "Faster-than-normal corn planting doesn't always hurts yields." Fortunately, a colleague noticed the aberrant *s* on the word *hurt* and called my attention to it in time for me to make the correction.

Then there was the sentence in an e-mail I sent a member of one of my editorial boards seeking information about a discrepancy in my record of manuscripts received. I revised the message several times before I sent it, which caused the problem and introduced the error.

- The first version of one sentence read, "According to my records, you received one manuscript each in January, February, April, and May."
- The final version of that sentence was supposed to read, "Those records show one manuscript received each in January, February, April, and May."

I read the final version quickly and sent the e-mail off. Sometime later I read it again and noticed the sentence contained the word *received* twice: "Those records show you received one manuscript received each in January, February, April, and May."

Here's a similar sentence error from another self-study course I wrote. In this sentence the aberrant phrase is *in the at the*: "Solid soybean seeding boosts yields when there is ample moisture available in the at the soil at the correct time." I cannot remember the earlier version of the sentence. Whatever it was, though, I failed to delete the "at the" when I revised it. I caught and corrected this error after the course was in type but before it was published.

So, do we keep them?

If word processors can insert such gross and nearly undetectable errors, are they worth their bother and price? J.D. Oster of the University of California-Riverside answers that question this way:

"I get frustrated at times with how these programs can create perfect mistakes. I calm down quickly when this occurs, however. Why? When computers were not available, my rate of writing was something in the order of one page per week. Now it may not be completely correct, or written as well as some would like, but the rate is up to several pages per day. Still, some may say the old rate was better for all concerned."

How to correct the errors

The simplest way to keep from inserting new errors into a manuscript is to refrain from correcting old errors and from polishing less-than-perfect prose. Of course, nobody would recommend that solution.

More difficult would be to perfectly proofread everything before you submit it. Even if you did not make your corrections and revisions perfectly, you would catch and correct all the errors described here.

Still more difficult would be to write so perfectly that you never make an error and, on those extremely rare occasions when you do, correct them perfectly.

More reasonably, do the best job you can when you first write the piece and the best job you can when you make alterations. Next, proofread as well as you can, but do not immediately submit your manuscript. Each of my errors reported above occurred when I made the change and almost immediately sent off the piece of writing. When I wait a day or two (or if that is not possible, at least an hour or two) before I send the writing off, and proofread it one last time, I generally catch my aberrant word or letter errors.

Still greater success occurs when I ask another person to read my final version before I send it off. (In that regard, I thank the friend who caught just such an error in the manuscript for this column. The sentence from three paragraphs earlier read, "Even if you did not make your corrections and revisions perfectly, you would catch and correct the all the errors described here." It would have been *really* embarrassing to have that error make it all the way into print.)

It is far better for my ego to have a friend call an error to my attention than to see it in cold hard print too late to do anything about it. Obviously I do not bother people who are themselves busy by asking them to help me proofread every little memo and letter I write. But I try to do this with anything major.

Chapter 5

A few specifics

As mentioned at the beginning of the previous chapter, I hope you have noted the gradual progression. I will get to specifics of grammar and style, but philosophically I believe it is more important to know what you are doing before you get too technical about how to do it. So this chapter moves a little closer to the nitty-gritty details, but asks some important questions and points out some important information along the way.

The first article, from August 1999, provides a positive response to the final column in the previous chapter. Sure, as the previous article stated, computers can cause problems. But as this next column demonstrates, a computer can allow you to become a far better writer than you ever imagined you could become.

It's a computer, not a typewriter

I chuckled once as I watched a typist use a computer for the first time, and hit the *enter* key at the end of each line as though it were a carriage return. I didn't chuckle too loudly, though. I probably would have done the same thing my first time, except that someone told me before I started that a computer will automatically wrap lines.

Computers can do a great deal more than merely wrap lines. I'll mention a few of the things I have my computers do for me.

I have noticed two distinct groups of computer users. One group uses the computer for everything. [I liked to joke about one person I knew by saying she had already written the software, and as soon as she could find the right hardware she intended to use her computer to brush her teeth.] The second group (people like myself) don't put anything onto a computer unless they find an actual benefit for doing so. Thus, the suggestions in this column, in my opinion, are those with a practical benefit. If you want more, look up books with titles like *The PC is Not a Typewriter*.

Search and replace

Have you ever used your search tool without automatically replacing something? As I was wrapping up a recent article I wondered how many passive verbs it contained. So I scanned the paper with the search tool looking for (one at a time) *is*, *are*, *was*, and *were*. When it stopped on one of those words I read the whole sentence to see if was the best possible verb, and rewrote about 75% of those sentences to replace the passive verb with an active one.

Want to verify that each reference is cited in the text? Search for an author name, or any other item (such as an opening parenthesis) likely to appear in a citation. Use the search tool to make certain you used everything else correctly each time. See also the next section.

Bookmark

I often use the *bookmark* feature during those searches. Particularly when I'm editing, if I come across something not quite right (say the word *nitrogen* instead of its abbreviation), I'll stop what I'm doing and put a bookmark there. Then I open the search and replace and put "nitrogen" in the search line and "N" in the replace line. I can then accurately scan the entire paper in only a few seconds and make all appropriate corrections. Then I go back to my bookmark and pick up where I had left off.

I'll use that sequence a dozen or more times as I'm editing a paper to catch and correct aberrant spellings, punctuation errors, capitalization errors, and many other such problems.

Quick correct

The common use for this tool is to correct those words that your fingers have just never learned to spell. Write the word once the way you usually do in the *replace* section of QuickCorrect, and once again (correctly) in the *with* section. Assuming your spelling error is consistent, you will never again have the word misspelled in anything you write on that computer.

Here's an even more valuable use for QuickCorrect. I often need to write particular groups of words or characters, like *American Society of Agronomy*, *Crop Science Society of America*, *Soil Science Society of America*, and *ASA–CSSA–SSSA*. In earlier computers I created macros to write them. QuickCorrect does this much more easily.

To get each of those phrases in the previous paragraph, all I had to do was type three characters. When I want *American Society of Agronomy*, I type * and a1. As soon as I hit the space bar the computer replaces those three characters to the full phrase. (The *and* in the previous sentence kept the machine from making the conversion there, where I did not want it.) When I want ASA–CSSA–SSSA, I type * and a2.

The way I originally programmed it, I got the desired phrase when I entered a1 (or whatever). That worked fine until I was preparing a table that had a1, a2, and several other items that were also in my QuickCorrect file. My table quickly filled with a host of unwanted words, symbols, and phrases. So I added the * before each of those entries in the file.

Two more QuickCorrects I use are * and WRL to give me *William R. Luellen*, and * and edit to give me *Managing Editor*. As you can see, when I do not use the * I can continue to use those two items as they are.

A Headquarters assistant editor has a number of phrases in her QuickCorrect file, including, "Please shorten your abstract to 250 words."

If you have items you often need to write (that do not require the complicated key strokes only a macro can handle), I'd urge you to create a few QuickCorrect entries and make your computer do some of the work.

In other cases, you might want to shut off or override other portions of your QuickCorrect. If yours (as many are) is programmed to convert (c) to ©, you may want to eliminate or modify that (perhaps with an *) so you don't get the copyright symbol every time you try to create a list of items.

I had a similar problem every time I tried to create a list within a sentence, using our style, which calls for (i), (ii), etc. Each time I would type (i), the computer would change it to (I), because it was programmed to always capitalize the single letter i. I corrected that by adding a line in my QuickCorrect to tell the stupid machine, "Replace (i) with (i)."

Windows

By this subhead I mean the feature, not the brand of software. When I'm editing a paper I'll nearly always have at least two windows open, and often four to six. I'll have the paper in one window, and in another window I write the questions I'll need to ask the author. I'll copy the paper's reference section to a third window, where I can work on it without accidentally affecting the original paper.

If I come across anything that needs rewriting, I'll copy it into another window and work on it there, with no danger of inadvertently damaging the original. When I have the problem fixed, I'll copy it back into the original and only then delete the item it replaces.

Toward the end of my editing, I'll place a template of the query letter in yet another window, and transfer information to it from all the other windows I've used for that paper (the paper's title and registration number and the corresponding author's name and address from the manuscript window; and specific questions, revised abstract, reworked sentences, etc. from my work window or windows).

I don't use quite as many windows when I write, but I'll often open a second window where I can revise a troublesome sentence or passage without the risk of messing up the rest of the paper. And, of course, I develop all tables and figures in their own windows and then either copy them into the main paper or save them as separate files.

Cut or copy and paste

As I prepared this column I contacted several people by e-mail to ask for their computer-use tips. I wrote a full address only once with the first name, then copied the @ and everything after it. For the people at Headquarters, for example, all I needed to write new was their first initial and last name. When I hit [control + v] I automatically got "@agronomy.org, " after that name. Note that I even included a comma and a space.

I've also used this technique to copy and paste all or part of position titles. Say I have a list of officers. I might copy *President* or *ASA–CSSA–SSSA* (created via the QuickCorrect method mentioned above), and then paste it after each subsequent entry.

Button bars

Button bars can be extremely valuable, especially if you learn how to customize them. I have them on three sides of the screen of my computer at Headquarters. One button gets me to the special characters some manuscripts need, such as diacritical marks for letters, footnote symbols for tables, and specialized dashes and other symbols. Another brings me to the type fonts. Two more buttons change the case of letters or words

I've highlighted, and another opens the tool that counts the number of words in anything (so I can immediately tell if an abstract is too long or calculate production charges). Still other buttons let me create an envelope, change page numbers, change line spacings, create tables, and do all the rest of the things I need to do.

On my home computer I do far more writing than editing, so the button bar on that computer reflects that difference. But I still have enough buttons on my home computer to allow me to work on anything I need to for the Society. [Update: After leaving ASA–CSSA–SSSA to found Fine-Tuning Your Writing, LLC., my computer work screen is far more complicated than anything I ever used at my old position. I have two rows of button bars at the bottom, one on the left, and at least one at the top all the time—all heavily customized. And I frequently will call up one or two additional bars at the top.]

Macros

Depending on the work you do, you might want to use many macros, or none. When most of my correspondence was by regular mail I used 10 macros to help me write it. Six macros dated the letter, added the inside address, and the greeting to the three editors-in-chief and three journal editors I corresponded with regularly. Four more macros added the closing, identifying me as the managing editor of one journal or all three. And I had buttons on one of my button bar to call up each one.

Your professional editors at Headquarters use several macros to enter typesetting codes into each manuscript as a part of preparing the electronic copy for our typesetter. They also have several other macros that have automated several other routine procedures on manuscripts.

Other

Depending on your hardware and software you use and the work you do, the *reveal codes* feature might range from essential to darn nice to not needed. [Unfortunately, MSWord does not have this capability, which can often drive an editor or even a serious author right up the nearest wall. See the first article in Chapter 16 for more on this.] I find *reveal codes* essential when I edit using WordPerfect 8 for Windows. I often call on it also when I'm writing at the office or at home.

I will not write much about spell checkers here because I think most of you are aware of their advantages and drawbacks. One drawback, if you add words to your supplementary dictionary, make sure you spell those words correctly. It appeared from one paper that an author had added both "fesuce" and "fescue" to his. Assistant Editor Gundega Korsts suggests that you check and, if necessary, prune your spell checker's supplementary dictionary from time to time.

I have found that grammar checkers are not much help, except in a few unusual situations. I used one a few weeks ago, but it was the first time it in a couple years. [And not again in the year or so since those words were written.]

To you Mac users who are wondering why I mentioned only PC tools. The reasons are because (i) the PC is the only computer I use and (ii) this column is already almost twice as long as normal.

* * *

As with many of the later columns in the original series, the following one was prompted by a question from a reader. This was first published in July 1998:

Some reference books that work

People sometimes ask me to recommend grammar and writing books; more often, they ask which books I use myself.

Interestingly, I rarely consult a grammar book, relying instead on language usage books. Two are considered absolutely the best in the field. I generally go to them in the order listed here:

A Dictionary of Modern English Usage, by H.W. Fowler, revised and edited by Sir Ernest Gowers, Oxford University Press. I bought my copy in 1965, just after it was published. A third edition was brought out within the past year. I have not bought a copy, but probably will soon.

While every writer (I've ever known, anyway) considers this the definitive source and refers to it as simply "Fowler," one must use it with some caution because it presents the British version of the language. Although I always go to Fowler first, I often find a more complete answer for my purposes in *The Careful Writer, A Modern Guide to English Usage*, by Theodore M. Bernstein, Atheneum. The particular version I have was published in 1965. It has been updated several times since and is available in most bookstores. Bernstein is one of the editors of the *New York Times*, so he presents not only the best of American English, he approaches it from the practical side.

[Update: I heartily recommend this book if you can find it. I could not find it when I searched for it after my retirement from ASA–CSSA–SSSA, but found another by the same author, titled *Dos, Don'ts & Maybes of English Usage*, 1977 Gramercy Books, New York city. This book is similar to *The Careful Writer* and just about as good.]

Another classic that virtually every writer refers to is *Elements of Style*, by William Strunk, Jr. and E.B. White (the latter also wrote the children's classic *Charlotte's Web*). I have a copy of what is usually referred to as "Strunk and White" at home, but for some reason (probably because it is not in dictionary format) I have never gotten into the habit of using it for reference. I re-read it every now and then. My copy of this book is copyrighted in 1959 by Macmillan Company and is the 15th printing, printed in 1962. [Update, a new edition of this book was just released. I had not seen it as these words were written.]

The Council of Biology Editors a number of years ago published a book titled *Scientific Writing for Graduate Students*, which would have been more accurately titled without the final three words. It was as good a book on the subject as I have seen (except for the one you are now reading, of course). I am told that CBE has updated and retitled the book, but I did not see it in a recent CBE catalogue, so I can give no further information.

Similar to it, though (and excellent), is Robert Day's book *Scientific English: A guide for scientists and other professionals*, (1992 The Oryx Press, Phoenix, AZ).

Another is just out, *Scientific Papers and Presentations* by Martha Davis of the University of Arkansas (1997, Academic Press, Inc., San Diego, CA). I have not yet read it, but I respect its author, and judging by the chapter titles and subheads, it is extremely thorough.

Some other books, while somewhat off-beat, contain excellent information on writing. Probably the best of that category would be *Language on a Leash* by Bruce O. Boston (1988, Editorial Experts, Inc., Alexandria, VA), which contains a collection of essays on good writing.

The same firm, Editorial Experts, Inc., also publishes a monthly newsletter, *Editorial Eye*, which I referred to in the previous chapter. I can generally find at least one worthwhile article in each issue.

Other sources

On the rare occasions when the above books will not give me all the information I need, I am fortunate to be surrounded by people who also have worked with words most of their lives; three other managing editors and a half-dozen associate and assistant editors. One of assistant editor at headquarters is a walking encyclopedia of grammatical information. Fluent in several languages and able to read several more, she knows language structure and grammar better than anyone I have ever known. Also on the ASA-CSSA-SSSA Headquarters staff, the membership registrar is a former English and grammar teacher. I consult with her from time to time as well. [Unfortunately, I'm not as able to draw on them as readily since my retirement.]

If you work on a campus, I would encourage you to cultivate a working relationship with one or more members of the English or writing departments. They could do for you what my colleagues do [did] for me.

Were it not for these colleagues, I would probably have to search for a good grammar book myself for those few times when Fowler and/or Bernstein are unable to answer my questions. By the way, I find the grammar checker on my Word Perfect software barely useful for only the most rudimentary problems. I seldom bother to open it up anymore. I do, however, use the spell checker on everything I write.

The old and new of it

At least some have told me apologetically that they still rely on their old high school grammar book. If you do also, you are in good company and have no need to apologize. The way the language has been deteriorating in recent years, older is definitely better right now.

There is an exception to the previous statement. George D. Gopen of Duke University has brought out two books, one on writing and the other on grammar, that are excellent. If you would like a sort of a preview of his writing book, look up the paper, "The science of scientific writing," he and a colleague published in the November-December 1990 issue of American Scientist (80:550–558).

List of resources

Here is a list of the writing, grammar, and style books compiled by Susan Ernst, managing editor of three ASA–CSSA–SSSA journals:

The ACS Style Guide. A Manual for Authors and Editors—*Edited by Janet S. Dodd. American Chemical Society, 1155 16th Street, NW, Washington, DC 20036. 1986. 264 p. ISBN 0-8412-0943-X.*

Editors on Editing—*Edited by Gerald Gross. Grove Press, 196 W. Houston Street, New York, NY 10014. 1993.*

Effective Writing Strategies for Engineers and Scientists—*By Donald C. Woolston, Patricia A. Robinson, and Gisela Kutzbach. Lewis*

Publishers, 121 South Main Street, Chelsea, MI 48118. 1988. 181 p. Hardcover. ISBN 0-87371-099-1

How to Copyedit Scientific Books & Journals—*By Maeve O'Connor. ISI Press, 3501 Market Street, Philadelphia, PA 19104. 1986. 150 p. Softcover. ISBN 0-89495-064-9.*

How to Write & Publish a Scientific Paper, 3rd Edition—*By Robert A. Day. Oryx Press, 4041 North Central at Indian School Road, Phoenix, AZ 85012-3397. 1988. 211 p. Softcover. ISBN 0-89774-456-X.*

Mark My Words. Instruction and Practice in Proofreading—*By Peggy Smith. Editorial Experts, Inc., 85 S. Bragg Street, Alexandria, VA 22312. 1987. 154 p. (plus exercises and answer keys). Softcover. ISBN 0-935012-08-7.*

Scientific English. A Guide for Scientists and Other Professionals—*By Robert A. Day. Oryx Press, 4041 North Central at Indian School Road, Phoenix, AZ 85012-3397. 1992. 125 p. Softcover. ISBN 0-89774-722-4.*

Stet! Tricks of the Trade for Writers and Editors—*Edited by Bruce O. Boston. Editorial Experts, Inc., 85 S. Bragg Street, Alexandria, VA 22312. 1986. 310 p. Softcover. ISBN 0-935012-07-9.*

Writing in Academic Disciplines. Advances in Writing Research, Volume 2—*Edited by David A Joliffe. Ablex Publishing Company, 355 Chestnut Street, Norwood, NJ 07648. 1988. 258 p. Hardcover $67.50. ISBN 0-89391-434-7.*

* * *

Another problem is the reaction of some authors when they hear that long and rambling sentences are difficult to read. Instead of correctly solving the problem, they go too far the other way. I noticed this problems in some of the papers I edited at ASA–CSSA–SSSA, and wrote this column about it that was published in September 1992:

Concise does not always mean short

The best writing is concise. It uses no more words than necessary in each sentence, no more sentences than necessary in each paragraph, and no more paragraphs than necessary in each paper.

The words "than necessary" in each of those phrases are important.

In arbitrary efforts to shorten their writing, authors sometimes take out necessary words. The result is sentences or paragraphs that have been shortened to the point that they are cryptic and difficult to understand. In cases like these, a few more words in a sentence or an additional sentence or two in a paragraph can greatly improve the level of understanding.

Say authors write, "We decided not to run the X diagnostic test." Without further explanation the reader cannot tell whether the researchers were being lazy, frugal, or omitted the test for a scientifically valid reason.

Let's take another example, and say one year of an experiment had average rainfall, one was exceptionally wet, and the third was exceptionally dry. You might improve reader understanding if you refer, at least part of the time, to "the wet year," "the dry year," and "the normal year" instead of using the much more compact 1988, 1987, and 1989.

A few authors will drop words such as *an* and *the*. Sometimes they are correct. Other times one of these words can greatly improve reader understanding.

Most mechanical standards and guidelines will tell you that 33 words are far too many to have an understandable sentence. Those guidelines are often correct. The following 33-word sentence, though (adapted from a paper accepted for publication in one of our journals) is actually too short to be clearly understood: "Finley (1971) concluded that the overall yield improvement of a barley bulk hybrid population under natural selection is due to increases and decreases in the numbers of high-yielding and low-yielding lines, respectively."

To make this sentence understandable I increased its length to 35 words by adding seven words and deleting five: "The overall yield improvement of a barley bulk hybrid population under natural selection is due to increases in the numbers of high-yielding lines and corresponding decreases in the numbers of low-yielding lines (Finley, 1971)."

This 35-word sentence is concise; it contains no unnecessary words and cannot be shortened or made more simple than it is. Concise does not necessarily mean short, it means no longer than necessary.

Needless to say, a paper should not be made up only of sentences that are each 35 words long. Varying sentence length makes a more reader-friendly piece. If you analyze this piece, for instance, you will find that its longest sentence contains 58 words and its shortest contains only four.

* * *

Conciseness is not the only thing that is important, of course, so is the order in which you assemble the words, as this June 1998 column shows:

Word order sometimes influences meaning

This column was prompted by a letter from a member. Before I get to that letter, though, I want to state the premise that the place a particular word is positioned in a sentence can influence the meaning of that sentence, sometimes profoundly.

To demonstrate, I will give eight sentences that are identical except for the placement of the word, *only*. One of the sentences is illogical. Of the seven sentences that make sense, six have totally different meanings, given parenthetically after each sentence.

- Only he said he could do that research. (No one agreed with him.)
- He only said he could do that research. (But, he did not demonstrate it.)
- He said only he could do that research. (He's the world leader.)
- He said he only could do that research. (Two possible meanings: [i] If you emphasize the *do*, it means he could do the research, but not write the paper or do any other followup. [ii] If you emphasize the *he only*, it again means he's the world leader.)
- He said he could only do that research. (And he refused to work on anything else.)
- He said he could do only that research. (He has extremely limited abilities.)
- He said he could do that only research. (Illogical sentence.)
- He said he could do that research only. (Limited abilities, again.)

The importance of word order to the meaning of a sentence is one reason why I try to avoid the use of *respectively* in a sentence. Occasionally the best way to present information is via the use of that word. More often, the word is tacked onto the end of a sentence in a vain attempt to bring order out of chaos.

Consider this sentence [the name and location were changed to save embarrassing the author]: "In continuous corn plots in northeast Siberia, Gorbachev et al. (1997) measured nitrate N losses in tile drainage water that ranged from 58 to 108, 63 to 76, and 11 to 20 kg ha^{-1} for 1990, 1991, and 1992, respectively."

Presenting exactly the same information without the use of *respectively* takes 42 instead of 40 words, but is far more understandable, especially the first time through: "In continuous corn plots in northeast Siberia, Gorbachev et al. (1997) measured nitrate N losses in tile drainage water that ranged from 58 to 108 kg ha^{-1} for 1990, 63 to 76 for 1991, and 11 to 20 kg ha^{-1} for 1992."

With all of this information as prologue, now I can deal with a question from Philip Smethurst, of CSIRO Forestry and Forest Products and CRC Temperate Hardwood Forestry in Hobart, Australia:

"I have trouble knowing where to place the word *also* in a sentence. I have thought about its common usage, but I think it is often misplaced. I [also] have [also] searched [also] in several English usage guides [also] to no avail. I would appreciate your advice and/or direction to an appropriate reference."

Let me begin my answer to that question by substituting the word *also* for the word *only* in the eight sentences I used at the beginning.

- Also, he said he could do that research. (As an afterthought, he said . . . .)
- He also said he could do that research. ([i] He, among several others. [ii] Afterthought, again.)
- He said also he could do that research. ([i] Afterthought, again. [ii] He, among others, again.)
- He said he also could do that research. ([i] He is a man of many talents. [ii] He, among others, again.)
- He said he could also do that research. (A man of many talents, again.)
- He said he could do also that research. (Illogical)
- He said he could do that also research. (Illogical)
- He said he could do that research also. ([i] Afterthought, again; or [ii] Man of many talents, again.)

Of the six sentences that make sense there are really only three meanings; half as many as when *only* was the variable word. Further, the meaning of most of those sentences could go more than one way, depending on which word in the sentence is emphasized. Why is it that the placement of the word *also* does not make nearly as much of a difference as did the word *only* in the meaning of this sentence? I believe the difference is due to the meanings of the individual words. The word *also* is much more encompassing in its meaning, while *only* is much more restrictive.

Now let us consider the four sentences suggested by Dr. Smethurst:

- I also have searched in several English usage guides to no avail.

- I have also searched in several English usage guides to no avail.
- I have searched also in several English usage guides to no avail.
- I have searched in several English usage guides, also to no avail.

I did not append a meaning after these sentences because I see no difference in their intrinsic meanings, and only slight differences in their emphasis and connotation. For subjective and aesthetic reasons I prefer the second and fourth sentences of this group. The second emphasizes the thoroughness of his search, the fourth its futility.

Grammarians will probably disagree with me about the second one because it breaks up the verb phrase *have searched*, which is not a good idea. Some grammarians would say, in fact, it should never be done. If you have read many of these columns you know how uncomfortable I am with the words *always* and *never*. I agree it usually is not a good idea to break up verb phrases. (Note that I wrote *usually is not* instead of *is usually not* in the previous sentence.) If you want to be a stickler about following that rule, the first or even the third of Dr. Smethurst's sentences conveys the thoroughness of the search about as forcefully as does the second one.

Now let's return to my opening statement: "the place a particular word is positioned in a sentence can influence the meaning of that sentence, sometimes profoundly." Now maybe you can see why I placed the word *sometimes* before the word *profoundl*y. Let's see if we can figure out why.

Go back to the first set of sentences in this column and consider the first and third examples, which are remarkably similar in construction and remarkable different in meaning. The first has the phrase, "Only he said," the third has the phrase, "only he could." In each case the word *only* comes before the word *he* to make its meaning clear. But what it is being made even more clear than the pronoun *he* is the verb that comes after that pronoun.

To conclude, the placement of a word can influence the meaning of the sentence. Whether that influence will be subtle or profound depends as much on the meaning of the individual word as it does on its placement among the other words in that sentence.

* * *

Even if the sequence of words does not affect their meaning, it can certainly affect their clarity, as demonstrated in January 1994:

Sequence of words affects their clarity

A sentence with all the correct words will make no sense if those words are in the wrong order. By this I don't mean something as simple as writing "I paper going am write to a" instead of "I am going to write a paper." I refer to papers where *most* of the words are in order, but certain key ones are not.

For instance, the following sentence begins with half of a phrase, "In a study." It is difficult to figure out what is being studied, though, because the second half of that phrase doesn't come until much later in the sentence:

> "In a study using both F_1 and F_2 seed, Kaeppler and Rasmusson (1991) found mid-parent heterosis was 85% for F_1 and 35% for F_2 seed for alpha amylase activity in three six-rowed barley crosses."

After reading that sentence several times, I finally realized that the study in question was of alpha amylase activity. When I placed those words directly after "In a study," the phrase was then intact and the whole sentence became clear with no further changes:

> "In a study of alpha amylase activity, utilizing both F_1 and F_2 seed of three six-rowed barley crosses, Kaeppler and Rasmusson (1991) found mid-parent heterosis was 85% for F_1 and 35% for F_2 seed."

Similar to the previous sentence, where a phrase was split and widely separated, too great a distance between the subject of the sentence and its verb can also cause comprehension difficulties. Here is an example:

> "Lesser amounts of wax myrtle (*Myrica cerifera* L.), wiregrass (*Aristida stricta* Michaux), grapevine (*Vitis rotundifolia* Michaux), and red maple (*Acer rubrum* L.) were also present."

This sentence actually means, "Lesser amounts of [something] were also present." The problem is that the [something] is an 18-word-long phrase. Because that phrase could not be shortened, I needed to use some other technique to solve this problem. The technique I settled on was restructuring the subject and verb so they were backwards from the traditional word order. The beginning of the sentence (the rest was unchanged) then read this way:

> "Also present were lesser amounts of wax myrtle . . . ."

Another problem is that some phrases are stuck into a sentence without thought of where they would be the most logical. Usually these orphan phrases begin with the word "and." The out-of-place phrase in the next sentence is the final one: "and hills were spaced 61 cm apart."

> "Both experiments were planted in hill plots where a single plot consisted of five hills with five seeds per hill and hills were spaced 61 cm apart."

I moved the orphaned phrase to a more logical location:

> "Both experiments were planted in hill plots where a single plot consisted of five hills spaced 61 cm apart, with five seeds per hill."

Any sentence with the word *respectively* in it should be carefully considered; there is almost always a better way to state the thought. A sentence in a manuscript I was reading the other day had at least a half-dozen numbers scattered through it, and ended with the word respectively. I had no way of knowing how the numbers were supposed to be linked to each other by that final word. I made a guess and asked the author if I was correct. I was not. My guess and the confusion it implied, though, prompted the author to restructure the sentence to both clarify the phrase and eliminate the often-confusing word.

Aside from *respectively*, most words serve a legitimate purpose in a sentence. But monotony (or even unplanned humor) can come when the same legitimate word is used over and over. The next sentence, for instance, might be called a *different* sentence because the word *different*, or variations thereof, appears four times:

> "On average, reciprocal differences were not associated with differences in cytoplasm, as the differences between the six crosses in *H. vulgare* cytoplasm vs. the six crosses in *H. spontaneum* cytoplasm were not different ($P < 0.05$) in any instance."

That's simply too much use of an otherwise good word. I reshuffled and divided the sentence to eliminate half of them:

> "On average, reciprocal differences were not due do the cytoplasm of the two strains. The six crosses in *H. vulgare* cytoplasm were not in any instance different ($P < 0.05$) from the six crosses in *H. spontaneum* cytoplasm."

Developing a feel for detecting out-of-place phrases is not always easy. When I read something I have written, I know what I meant, and often infer that instead of what is actually written.

Asking a friend to read the piece for me does not always help, either. The friend asks, "What do you mean by this?" and I reply tersely, "What do you mean, 'What do I mean!' I couldn't make it any clearer!"

When I cool off, apologize to my friend, and take a more rational approach, though, I see what the person had been talking about, make the required changes, and the sentence is definitely clearer than the original.

Regardless of whether you find the misplaced words yourself or ask someone else to help you find them, the important thing is that you move them to their proper location before anyone else finds them.

* * *

I am always on the lookout for ways to demonstrate particular points about writing. The need to shorten an abstract for an author provided one such opportunity, and serves as an excellent counterpoint to the one immediately above. This next one is from January 1997:

Weeding out your wordiness

The professional editing staff at Headquarters normally does not rewrite papers for authors. We had an unusual situation recently where an author was unavailable, a deadline was at hand, and we had an abstract that contained 296 words. After some brief discussions within the staff, it was decided that I would shorten the abstract to the limit of 250 words, send it to the typesetter, and hope the author, who would again become available after the abstract was in type, would approve my changes.

Because I was not about to pay a typesetter to set my version of the abstract, and then throw that away and pay the typesetter again to set another version the author preferred, I wanted to come up with a document that the author would approve with few or no changes. I knew, therefore, that I could not simply rewrite the abstract in my own words.

This made for an interesting challenge, to delete superfluous words without altering either the meaning or the character of the abstract. I ended up needing to substitute only one word and alter a couple others, shortening the original 296-word abstract to 253 words.

After I sent the shortened abstract off to the typesetter I decided the revisions might be interesting to others, so will detail them in this column. I hope after you read this you will be able to get a better feel for which words are necessary and which are unnecessary in a paper, and also perhaps the difference between *unnecessary but nice* and *unnecessary*, period.

To shield the author's identity I eliminated all references to the crop and location and will repeat only pertinent parts of some sentences and leave other sentences out entirely. The only sentences I will quote in their entirety are those that were eliminated. Where some reference to the crop was necessary, I did so only vaguely within brackets.

Original sentence: "Human population growth is placing ever-increasing demands on the world's [one type of crop]."

My change: "Human population growth places increasing demands on the world's . . . ."

Explanation: Using *places* instead of *is placing ever* not only shortens the phrase, it also changes a passive verb to an active one. The phrase is still in the author's words, though; I would have written, "Human population growth increases demands on . . . ."

I thought long and hard before deciding not to delete the word *human*, at the beginning of the sentence. While I do not believe the sentence would be misconstrued without it, the word adds some meaning to the sentence.

Another original sentence, with cross-outs to show my deletions: "Sustaining ~~a variety of values in~~ these [crops] will depend to a considerable extent ~~, though not exclusively,~~ on sustaining soil structure and function."

Explanation: Although it may not be clear without the title and the rest of the abstract, the deleted first fragment is redundant, as is the second phrase.

Original: "[until now] concerns about sustaining [crop] production focused largely on soil fertility, but recent interest ~~in sustainability~~ has focused attention on the plant and animal community and on 'biodiversity.'"

Explanation: The deleted words were redundant in this sentence.

Original: "Neither of these ~~disciplinary~~ approaches ~~to the evaluation of sustainability issues~~ are satisfactory on their own."

Explanation: The stricken words were redundant with the previous sentence.

Original: "They should be replaced with an ecosystem analysis approach that includes ~~both~~ the biotic and abiotic components ~~of the ecosystem~~."

Explanation: It really is not necessary to have the word "both" in front of the two items, and *an ecosystem analysis of the ecosystem* is redundant.

Original: "~~The concept of [the commodity] productivity is multifaceted.~~"

Explanation: There is nothing terribly wrong with this sentence and it shows a bit of the author's style and personality. I would not have deleted it if the abstract was within the space limit. This sentence was deletable, though, because it really says nothing; the concept of virtually all productivity is multifaceted.

Original: "~~However~~, the most fundamental issue in . . . ."

Explanation: Elimination of the previous sentence eliminates the need for the word *however* in this sentence. An added bonus (for me), I do not like to begin sentences of this sort with the word *however*. If I needed to use *however* in this sentence, I would have written it, "The most fundamental issue in . . . , however, ..."

Original: "Understanding the sustainability of site productivity requires knowledge of ~~the determinants of production ecology:~~ the determinants of leaf area, and its photosynthetic efficiency, and of the allocation of net production."

Explanation: This was a close call, but it seemed redundant to me.

Original: "Key factors that determine production ecology . . . ."

Revised: "Keys that determine production ecology . . . ."

Explanation: Another example of what I have written about many times; the word *factor* can usually be deleted from sentences such as this and never be missed. (Obviously you cannot eliminate that word from a sentence about factoring an equation.)

Original: "In the long-term absence of . . . ."

Revised: "Without . . . ."

Explanation: This was only total replacement of a phrase. I did it because the word does essentially all of the work of the six it replaced.

In all, I eliminated 43 words from the abstract, not one of which was essential for the presentation of the information. Yet the abstract still sounded as though it was written by the original author, not by me. My evidence to support the previous statement? The author accepted the shortened version without change.

* * *

In a similar vein, this one from November 1992:

Cumbersome constructions

Why is it that scientists will say to their colleagues, "We are trying to figure out how it works," but write, "The research was conducted to ascertain the function of the system"?

There seems to be a notion that it isn't proper scientific writing unless it is cumbersome and difficult to read. This notion ought to be relegated to the same place where we store our mechanical calculators, manual typewriters, and buggy whips.

Science is a complicated subject, but it need not be described in complicated sentences. The fact that science is complicated might be all the more reason to describe it simply and logically.

Consider the words of Stirling Moss, one of the greatest Gran Prix racing car drivers in history. When asked his secret of winning so many races, he said, "I always drove as slowly as possible." By that, he meant as slowly as possible to get him to the finish line ahead of all the other drivers. To drive faster than that—to win the race by a large margin—would stress his car unnecessarily and might cause it to break down before the end of the race.

How should sentences be constructed? Using the Stirling Moss analogy: As simply as possible. The remainder of this column will give examples of three sentence constructions that ought to be thrown onto the scrap heap.

Separated subject and verb

Consider the following sentence: "Heading dates when spikes from approximately 50% of the tillers had emerged were recorded." That sentence has 10 words between its subject (heading dates) and its verb (were recorded). That, unfortunately, is not unusual; we have seen sentences with as many as 25 words between the subject and verb. With such sentences, readers lose track of the subject by the time they get to the verb. Then they have to reread the sentence.

The easiest way to correct sentences with this problem is to simply place the verb directly after the subject. Thus, the sample sentence would read: "Heading dates were recorded when spikes had emerged from approximately 50% of the tillers." (Note that I also moved *had emerged* to directly after its subject, *spikes*.)

If you like, you could go one step further and change it from passive to active voice: "We recorded heading dates when spikes had emerged from approximately 50% of the tillers."

Backwards write don't

The subhead just above makes more sense if you reverse the order of its words. This is an illustration of a second awkward construction, sentences that are written backwards. The following is an example of such a sentence, with the location changed to protect the identity of the author: "Irrigation water at 25 mm per application at the first sign of moisture stress was applied at the Siberia location only." Aside from the word *Siberia*, every other word in that sentence was exactly as submitted. Here's the way I would have written it: "At the Siberia location we applied 25 mm of irrigation water at the first sign of each moisture stress."

Passive voice

The use of passive voice is so common in scientific writing that some scientists believe it is required. To prove this, I will first quote a passage from "Become a more successful author," a helpful essay we include in our journals from time to time: "Scientific writing contains far too much passive voice. Let's start moving away from it, as we tried to do in this piece. Regardless of what anybody tells you, it's okay to use first person in scientific literature. You don't have to say 'the research was conducted,' you may say 'we conducted the research.'"

As this document was wending its way through the review process, prior to its first publication, one of the reviewers asked, "Don't we need to change the by-laws before we can say this?"

Because it was a blind review, I was unable to ask that reviewer if that was an ironic comment or meant to be serious. But passive voice has never been required in ASA–CSSA–SSSA by-laws (or anyplace else that I've ever been able to find).

I can think of only one objective reason for using passive voice; to soften an uncomfortable passage. Biological scientists usually write "the test animals were euthanized" instead of "I killed the test animals." A subjective reason for using passive voice would be to inject variation in the writing, as this sentence does.

[See Chapter 13 for a thorough discussion of passive voice.]

Conclusion

You won't impress readers with your use of cumbersome sentences, you will frustrate them with something that is difficult to understand. The best way to impress your colleagues is to convey your research so clearly and forcefully that they can easily understand it. And never forget that the readers of scientific journal papers include beginning graduate students, bright undergraduate students, and scientists at all levels whose first language is something other than English.

* * *

When you find a problem in your writing, do you think it takes a degree in English or journalism to correct it? Think again. Some corrections are amazingly easy to perform, as shown in November 1997:

Complex problems, simple solutions

Sometimes the difference between an extremely difficult-to-read sentence and an easy-to-read sentence is smaller than you might realize. For example, when an author constructs a sentence with a large distance between two components that should have been close together, the simple movement of a word or two can often be the only change necessary.

Move the verb

Gaps between subject and verb are the most common of these errors. To fix it, either move the verb or create a new verb earlier in the sentence. Here is an example of a sentence with that problem:

"A solution to the horrendously complicated problem of how to sequence the various components of this fully integrated in-vitro system has been discovered."

Here is the same sentence with the verb moved from the end of the sentence to just after the subject:

"A solution has been discovered to the horrendously complicated problem of how to sequence the various components of this fully integrated in-vitro system."

You could also convert the sentence to active voice:

"We discovered a solution to the horrendously complicated problem of how to sequence the various components of this fully integrated in-vitro system." If you do not like the word *we*, substitute: "A team of scientists at State Tech discovered a solution to the horrendously complicated problem of how to sequence the various components of this fully integrated in-vitro system."

Release the verb

The next example is more complicated. It has a noun and a verb early in the sentence and close to each other, but it also has words later in the sentence that prevent them from acting as the sentence's subject and verb:

"Buctril can be tank mixed with Butyrac to improve control when weeds in the mustard or smartweed family (which are sensitive to Buctril), and pigweed (which is more sensitive to Butyrac) are present."

To fix it, all I needed to do was change the later words, allowing the noun and verb already there to function as subject and verb:

"Buctril can be tank mixed with Butyrac to improve control of weeds in the mustard or smartweed family (which are sensitive to Buctril), and pigweed (which is more sensitive to Butyrac)."

An active-voice alternative to this sentence is even simpler:

"Tank mix Buctril with Butyrac to improve control of weeds in the mustard or smartweed family (which are sensitive to Buctril), and pigweed (which is more sensitive to Butyrac)."

Simplified even more:

"Tank mix Buctril with Butyrac to improve control of mustards or smartweed (which are sensitive to Buctril), and pigweed (which is more sensitive to Butyrac)."

Unsplit the verb

The final example has a verb phrase widely split:

"Recently, automated systems for estimating plant spacing variability have been developed."

There are a number of ways this sentence can be fixed. I chose to move the parts of the verb phrase together at the end of the sentence:

"Automated systems for estimating plant spacing variability have recently been developed."

Here are two alternate solutions for those who do not like the still-split verb phrase:

"Automated systems for estimating plant spacing variability recently have been developed." "Automated systems for estimating plant spacing variability have been developed recently."

An entirely different solution is more along the lines of my first example in this column:

"Automated systems have recently been developed that estimate plant spacing variability."

Here's a version for those who prefer active voice:

"Engineers here developed automated systems for estimating plant spacing variability."

And two still simpler versions:

"Plant spacing variability can now be estimated with automated systems." "Automated systems can estimate plant spacing variability."

How much to change

In this column I tried to show two to three levels of change to each problem. The first level in each case was extremely simple, retaining all or essentially all the original words, but rearranging them in a more logical order. The second level involved a minor rewriting to convert the sentence from passive to active voice. The third level, shown in the final two examples, involved a more complex rewriting of the sentence.

Any of these levels could be used by an author to improve sentences in his or her own paper. If you are an editor working on someone else's paper, though, I would recommend using only the first level of change in each case, which keeps the sentence in the author's own words.

* * *

About a year later, October 1998, I wrote a similar column:

More than one way to correct a problem

In these columns I have always encouraged you to write freely, yet I generally show only one way to correct a particular error or poor sentence construction. I'm going to try something different this time, and show at least two solutions to each problem.

In most cases, each solution will be roughly equal in acceptability, so I will not comment as to which solution I prefer. (In a couple cases where I offer more than two solutions, I may suggest that one may be less acceptable than the others, but I will always leave at least two equally acceptable ones for you to choose from.) As to which of the recommended solutions you should use, choose whichever you like best.

Run-on sentence

Incorrect sentence: "Precision and accuracy of load cell measurements were assessed using mass balance calculations and indicated agreement within a few tenths of a percent total mass."

Solution 1, make it two sentences: "Precision and accuracy of load cell measurements were assessed using mass balance calculations. They indicated agreement within a few tenths of a percent total mass."

Solution 2, make a dependent clause: "Precision and accuracy of load cell measurements, assessed using mass balance calculations, indicated agreement within a few tenths of a percent total mass."

Solution 3, make an independent clause: "Precision and accuracy of load cell measurements, which were assessed using mass balance calculations, indicated agreement within a few tenths of a percent total mass."

Unusual spellings

From a letter:

Recently I have encountered a dilemma in the writing of my thesis concerning a person's name such as van Ginkle or de Vasco, which is not capitalized (the *v* or *d*). When the name is used at the beginning of a sentence, is it capitalized? This would seem to be a misspelling of the authors name, though convention would rule that a sentence always begins with the first word capitalized.

— *Bradley L. Rauh*
University of Missouri
Columbia, MO

My reply:

About the only thing remotely definitive regarding the handling of such names at the beginning of a sentence is the *Encyclopedia Britannica*, which begins a sentence on Vasco de Gama with "De Gama . . . ." Therefore, it must be all right to capitalize the name at the beginning of a sentence. Practically speaking, though, you would be better off to try to work your way around it in one of two ways.

Solution 1, insert a title: "Dr. van Ginkle says . . . ." "Mrs. de Vasco says . . . ." "President van Ginkle says . . . ."

Solution 2, move the name inside: "According to de Vasco, ..."

Solution 3, make no change: In reference sections, membership lists and other places of that sort (in other words, anything that is not a standard sentence), we at ASA-CSSA-SSSA and most other publishers go with the lower case so as not to misspell the name.

Passive voice

Incorrect phrase: "The test was unable to show . . . ."

Solution 1: "The test did not show . . . ."
Solution 2: "The test failed to show . . . ."
Solution 3: "We could not determine from the test . . . ."

Cumbersome constructions

Awkward sentences: "In the U.S. Midwest, during the frequent high intensity storms that occur in spring on wet soils, localized surface ponding is likely to occur. Earthworm burrows open at the soil surface will be available for conducting this ponded water."

Solution 1: "In the U.S. Midwest, localized surface ponding is likely on wet soils during frequent high-intensity spring storms. Earthworm burrows open at the soil surface conduct this ponded water."

Solution 2: "Localized surface ponding is likely on wet soils during frequent high-intensity spring storms in the U.S. Midwest. Earthworm burrows open at the soil surface conduct this ponded water."

Gender neutrality

Incorrect sentence: "A scientist who uses this device should be careful to make certain his eyes are protected from possible damage from the chemicals contained in the device."

Solution 1, make it plural: "Scientists who use this device should be careful to make certain their eyes are protected from possible damage from the chemicals contained in the device."

Solution 2, use both pronouns: "A scientist who uses this device should be careful to make certain his or her eyes are protected from possible damage from the chemicals contained in the device."

Solution 3, rewrite the sentence into second person format: "Be careful when you use this device to make certain your eyes are protected from possible damage from the chemicals inside it."

Solution 4, use second person and also simplify the sentence: "Make certain you protect your eyes from the chemicals inside this device."

Solution 5, violate a grammar rule: "A scientist who uses this device should be careful to make certain their eyes are protected from possible damage from the chemicals contained in the device." (I don't personally like and do not use this solution, but have noticed it being used a great deal in recent years: "Everyone should get their income tax return in by April 15." "Everyone should exercise their right to vote next Tuesday.")

* * *

Another thing scientists often need to do is cite manufacturers and mention products that they used. Similar to the previous two columns, there are some solutions to this often complex problem that are far more palatable than others. From December 1992:

Citing manufacturers, using trade names

The basic premise behind any scientific paper is to provide sufficient information so a competent scientist can repeat the experiment. If you keep that thought in mind, most requirements outlined in the your publisher's style manual fall logically into place.

Manufacturers

Few ASA–CSSA–SSSA style rules have caused more confusion than the requirement to describe the materials used and to identify the manufacturer of commercial products. We are continually asked whether it is necessary to identify the manufacturers of every product.

In reply, I ask the author to consider whether the particular product is essential to the outcome of the research. For instance, if any test tube will work, an author need only say that the material was placed into a test tube. If, because of special circumstances, only a Teflon-lined or Pyrex (or whatever) test tube will work, then that must be stated. If that test tube is difficult to find, you have to tell the reader where to find it.

If only one infrared thermometer is accurate enough for the measurements that must be taken, then specify it. If not, you don't have to write any more than, "Temperatures were measured with an infrared thermometer held . . ." and give the specifications, such height above the crop canopy, angle of aim, etc.

If you write that you examined something under a binocular microscope, we won't ask you who made the microscope. If you say you used a Luellen no. 5041 binocular microscope (or any other brand that is not well known) Headquarters editors will assume that your research required using that brand and model and will ask you to tell other researchers where they can get it.

In other words, the editors in Madison (and most publishing houses) take their cue from the paper itself. If you give us a false clue, don't blame us for asking the logical question.

Sometimes a particular problem arises. A recent situation involved a piece of equipment that was essential to the research project, but was no longer available and its manufacturer was no longer in business. The author had three options:

1. Identify the product and the manufacturer anyway, even though that information would have been useless to the reader.
2. Identify the product and the manufacturer; explain that it is no longer available; say that it is similar to a product that is still available; and give the manufacturer's name and address.
3. Identify only the current product and manufacturer.

We chose the third route, with a sentence that read something like, "The product we used is no longer available, but is similar to the no. 3 Widget built by the North American Widget Works, Bayonne, NJ."

Perhaps you noticed that we do not always give an address of the manufacturer. A few products that are so widely known or so universally available that it is not necessary to tell scientists where to find them. Whatman filter paper, IBM and Apple computers, and Zeiss microscopes are examples. If an author includes an address of these manufacturers we do not delete it, but we do not ask the author to supply it.

The same is true for the more popular brands of seed. If you want some Pioneer seed, you probably will contact your local Pioneer dealer and not write to Johnston, IA, so we do not require you to include its address (but neither will we delete it if you include it). We follow the same rule for other major seed manufacturers. Companies that do not have such extensive distribution networks should be identified by their headquarters address.

Trade names

Use of trade names poses problems similar to those mentioned above, but with an added dimension; the use of the name is prescribed by law. For instance, you cannot plant a seed in a styrofoam cup because there is no such thing; Dow Corning, which holds the trademark on the name Styrofoam, does not make—or allow anyone else to make—a cup that it calls a Styrofoam cup. If we allow you to misuse its name in our journal, Dow will send us a forceful letter. You may write that you planted seed in a foam plastic cup or a cup made of foamed styrene plastic.

Trade names cannot be changed in any way. You cannot say, for instance, that Coke's share of the worldwide cola market is larger than Pepsi's share, because trade names can neither be made into a possessive or pluralized. You would have to restructure that sentence so the product names appear as Coke and Pepsi.

There are many more rules involving trade names. Professional editors keep track of them for you so you don't have to. This allows you to concentrate, instead, on your scientific work.

* * *

I went into this subject in deeper detail in two articles in the July 2000 issue of my newsletter. I conclude this chapter with those articles:

Trade names, their use and misuse

Few words cause more difficulties and headaches for authors, editors, and publishers than those that are trademarked. Each of those groups—authors, editors, and publishers—bring not only their own viewpoint to the discussion, but also carry additional burdens to the table.

The weed scientist whose work was partially sponsored by a particular manufacturer will want to say that he applied Treflan (or whatever the name of the specific product might be) to the weeds. The non-profit organization that publishes the journal in which the paper was published will want to change that to trifluaralin, to avoid the appearance of favoring a particular brand of herbicide above all others with the same chemistry. The government agency who employs the weed scientist will have its own say in the matter.

Happily, those matters can usually be resolved. Once that matter is resolved, though, subsequent steps become even more tricky because they involve necessary information, legal technicalities, scientific integrity, and a large dose of common sense. I will deal with three of these in this feature, and with the legalities in the companion article that follows.

How important is the brand name?

The style manuals of most publications require authors to describe the work in a paper in specific enough detail to allow a competent scientist to be able to duplicate the work. Scientist–authors want to comply, but are not always certain how precise they need to be in describing the work and how it was done. The question they need to ask themselves and people like me is, "How much of that information is necessary to duplicate the work?"

A paper I worked on recently contained the phrase, "We examined the leaves with a 10x hand lens." That is a perfect example of all that is needed. I'll explain why as I examine the matter in some depth.

Let's say that the particular hand lens those authors used was manufactured by Zeiss. If you were the author, would you have included that information, too? There are probably dozens of manufacturers making such lenses and selling them under scores of brand names. But with a product as simple as a hand lense, I suspect you will find little, if any, difference in the laboratory-quality lenses available to a typical scientist. Indeed, most standard hand lenses made for office use are likely similar in quality as well, and depending on the materials used to make the handle and the frame that holds the lens, many of them would also be suitable. So simply stating that a 10x hand lens was used is likely all the information a scientist would need to know to duplicate the research.

Now let's take that principle and move onto more complicated questions. Let's say the chemistry portion of your work was performed in glassware supplied by Edmunds Scientific Co. Should that name be mentioned in the paper? When authors ask me that question, I reply with a question of my own, "Is use of that particular product essential to the outcome of the work?"

In another words, if any old test tube will do, then I would say it is not important to mention that the one you happened to use came from Edmunds. All you would need to give would be its size.

Suppose, though, the glassware must be able to contain a certain amount of pressure or an unusually high temperature. Or suppose that the glassware must be lined with Teflon or some other special lining. Suppose that the glassware must have certain properties that are met only by Pyrex glass. Obviously that must be mentioned. Suppose only one manufacturer of scientific glassware makes the appropriate flasks, vessels, and tubing out of Pyrex glass? Then you have no choice but to mention that fact and give the name of the manufacturer and the address unless the manufacturer is well known.

Or, let's say you used a spreadsheet to analyze the results of your work; did it have an unusual set of properties required to handle your work, or will any spreadsheet work? If there is only one that will do, then you need to state that fact and tell both its name and who made it.

Will the spreadsheet you used run on any computer, or are there specific requirements for memory, speed, or whatever? If only a particular model number of a Dell or Gateway or Apple or whatever computer will handle the spreadsheet, then by all means let the people who are going to use it know. Must that computer have a Pentium III chip operating it, or will any high-capacity chip made since July of 1999 be sufficient?

If you use a fairly large soil sampler mounted in a truck instead of a traditional hand auger, first decide if you really needed to use that large a machine (or if it was simply the one that was most readily available at the time you took your samples). If you needed to use it, based on the number or depth of samples or nature of the soil, then mention it.

Then, as with the spreadsheet above, did it need to be mounted in the Ford F150 you used, or would a Ranger or Dakota also haveworked? If it had to be a full-size pickup, did it have to be a Ford (because of electrical or hydraulic hookups, or whatever) or would any brand do? Did it have to be a 4 by 4, or would a smaller, less powerful truck also fill the bill? The answers to all the above questions will determine the amount of detail you need to include in your paper.

How well to identify

Let's say the answer to all the above questions was yes, you did need the specific brand and model you used. Now, how thoroughly do you need to identify it? I'm sure you are aware that the style manuals of many publications say you need to give the name and address of the manufacturer. But you also need to use some common sense. If you needed some Whatman filter paper, you most likely would not contact the factory, you would get it from your local supplier.

Here's where things can get tricky, though. A product that is readily available in Urbana or Ames may not be so readily available in Manila or Cali. But for the most part, the well-known laboratory products are available fairly universally. In other words, if you need a Ford truck, you would contact your local dealer, not the factory in Dearborn, MI.

Now for the exceptions. In one paper I was working on recently, the author had made use of an experimental implement from a reasonably well-known manufacturer. I kept the name and address of that manufacturer in the paper because it was not likely that its distributors would have had access to that particular experimental implement.

In another recent paper the author gave both the exact specifications of an implement and listed the name and address of the manufacturer. That particular manufacturer had gone out of business years ago and it would have been nearly impossible for a reader of the journal paper to find another of these implements. So I deleted the name of the manufacturer, but retained the exact specifications. My reason for this was to assist the person who wanted to duplicate the work to find a similar piece of equipment.

Care and use of trade names

What do cellophane, aspirin, and nylon have in common? They all used to be brand names. What they also have in common is that (i) these names were misused by journalists and others like you and me, (ii) the manufacturers did not take steps to combat this misuse, and (iii) those manufacturers therefore lost their rights to those trade names.

What do Xerox, Rollerblade, Styrofoam, and Coke have in common? They are among a group of brand names whose owners are fighting hard to retain ownership and use of those names.

Because of what happened to the brand names mentioned in the first paragraph, today's manufacturers are much more careful and work much more diligently to protect their brand and trade names. If you misuse a trade or brand name in print you will likely get a polite but forceful letter from the legal department of the owner of that name.

So what is misuse and what is proper use of a brand or trade name? That answer, in part, depends on who is doing the answering. The legal department of the holder of the name will give you a precise and detailed set of instructions which, if followed, would make your text sound like something approximating an advertisement for the product.

Editors and people like me will tell you ways to meet the legal requirements in a more realistic way.

Technically, a brand name is an adjective. In other words, the attorneys will tell you that you cannot drink a Coca Cola, you must drink a Coca Cola brand cola soft drink. That's true, but you can also drink a soft drink manufactured by the Coca Cola Company. They will tell you they would like to see the name itself capitalized in the same way that they capitalize it — for instance, DICKEYjohn. They will also tell you that they would like to see the name followed by the superscripted "TM" (for trade mark) or circled "R" for "registered" symbol, or occasionally "SM" for service mark. Please note my specific use of the words "must" and "would like," in the previous sentence. They are important. Just because someone would like you to use something does not at all mean that you must.

For myself, I try to walk a middle road. I acknowledge the need to protect a trade name, but I also acknowledge the need to write material that does not sound like an advertisement. Here's what I mean:

First of all, there is no such thing as a Styrofoam cup, because the only company that owns the right to make products with the brand name

Styrofoam on them is the Dow Chemical Co., and it does not make cups with that brand name. So if you wrote in a paper that you started seeds in Styrofoam cups and the editor did not catch and correct that sentence, that editor, and possibly also you, will get a letter from Dow's legal department.

Trade mark holders also take offense when you write that some other product is "like" their brand-named product, because (in their way of thinking) there is nothing else in the world like their product. Therefore, writing, "a cup made of a foamed styrene plastic like Styrofoam" will also elicit a letter from the firm's attorney.

Leave out the offending word, *like*, though, and you will likely have no problem with, "a cup made of foamed styrene plastic." Not only is the phrase just given better in the eyes of lawyers, it is also more correct, anyway.

Dow makes insulation sheeting under the Styrofoam brand name, and you might think, then, that it would be all right to write, "insulation sheets made of Dow Styrofoam," but you would be wrong. The reason for that is because this phrasing might imply, in some people's minds, that other manufacturers also use the name—something on the order of "insulation sheets made of Acme Styrofoam."

Dow makes Styrofoam insulation sheeting. The best material for gluing that sheeting to something else is Styrofoam mastic, a glue with the consistency of well-chewed bubble gum.

I demonstrated two things in that previous paragraph. One is to show you once again how to use a brand name in the way the attorneys cannot object to. The second is to also show that the Styrofoam brand name (as one of many examples) is used on products other than things made of foamed styrene plastic. I most definitely would not want to take a sip of a soft drink (whether it was made by Coke or anyone else) out of a cup made of Styrofoam mastic.

One of my clients was giving me information to place into the company newsletter describing an upcoming remodeling of its offices. He said all interior doors would be covered with Formica. I asked, "Really? Or will you use laminated plastic sheeting?" He nodded, "You are correct, laminated plastic sheeting."

Most writers, when they first receive word from attorneys telling them the "proper" way to handle a trade name will get huffy and refuse to comply because they do not want their prose to sound like an advertisement. As I have just demonstrated, there are ways to legally use a trade name in a way that offends the sensibilities of no one.

Chapter 6

Punctuation and Grammar

My statement in the introduction of this book about how I try to be careful not to scare anyone who may have been traumatized by a grammar teacher was not meant to be in jest. In writing such as my newsletter, this book, and correspondence with clients I will use terms such as *subject* and *verb* from time to time, and occasionally use the term *object*, but I deliberately try not to use any other grammatical jargon if there is any possible way to get around it.

The reason for this is simple, If I'm going to insist that scientists refrain as much as possible from using their jargon when writing to nonscientists, I figure I should follow my own preaching and not use grammatical jargon when I'm writing for nongrammarians.

Besides, I've never liked most of the jargon used by grammarians and English teachers. I think it can sometimes get in the way of communication. That is not just an opinion. In the final article in this chapter, in fact; I will demonstrate an example of that happening.

This is not to say that I do not use correct grammar in my columns, newsletter, and other writing. I use it all the time. This is also not to say that I do not teach the proper use of good grammar, because I do. But I teach it as much as possible without the jargon.

I call the technique I use to teach grammar *subliminal seduction*. I'll write something like, "You don't need to remember any arcane grammar rules, all you really need to remember is . . . ." And then I present the grammar rule, but in a way so it doesn't sound like it's a grammar rule.

Be that as it may, from time to time I have needed to get specific, and rarely even use some of the jargon in writing about some grammar rules. I've placed those columns into this chapter not in hopes that you will skip it, but specifically for those who think a book on writing must contain such things. I begin, in fact, with some grammar questions that were asked by readers and answered in a column first published in April 1997:

Questions I've been asked about grammar

You have written that you do not like the use of *however* at the beginning of a sentence. I have style manuals that agree, but I have never understood that. The use of *however* alerts the reader that the sentence will present a different view from the one preceding it, which is, I believe, a useful device. The word *however* is a shorter version of "On the other hand" and a longer version of "But." Placing *however* at the sentence beginning is at least as useful, or more so, than putting it into the middle of the sentence.

This construction seems to me to be perfectly clear, not wordy, and helpful to the reader. Further, it is commonly used in conversation. Why do you object to it?

Arnold P. Appleby
Oregon State University

A complete answer to that question would require more of a history lesson than I care to give and you likely would care to read. There were good reasons for the rule when it was established, but English continues to evolve. Today, many respected writers begin sentences with *however*. One of these is Robert Day in his book *Scientific English: a guide for scientists and other professionals*, (1992 The Oryx Press, Phoenix, AZ).

Many editors, including some on your Headquarters staff, believe this once-incorrect use of the word *however* should now be considered correct. While I do not entirely agree, and prefer to use the simpler *but* myself, I appreciate the reasoning of those who use this construction. As of this writing I will no longer fight those who want to use it.

—WRL

What is the basis for the current trend toward hyphenating *ly* adverbs in compound adjectives? Example: "a highly-porous soil."

Oxford and Webster and my various style manuals (including E.B.

White) tell me the hyphen is redundant and inappropriate as *ly* indicates the adverb modifies the adjacent word.

Has there been a rule change or is this another something that comes with word processors and their grammar checkers?

R.S. (Bob) Loomis
U.C. Davis

If large numbers of people were to start robbing convenience stores that would not be a trend, it would simply be wrong. Many people today misspell *angle* as *angel*. That also is not a trend, it is wrong.

And although many people today hyphenate adverbs, this also is not a trend, it is wrong. As Bob Dole might put it, "I know it, you know it, and our professional editors know it." Too many American people do not know it, though, and the error happens frequently (although not as often in ASA–CSSA–SSSA publications as in others of lesser esteem).

Your comment about grammar checkers may have some merit. A number of words (*early* comes immediately to mind) happen to end in *ly* but are not an adverbs. Since "early-morning sampling" is correct, maybe the dumb machine thinks "highly-porous soil" is also correct. [A complete discussion on the use of hyphens begins on page 138.]

Speaking of grammar checkers, one objected to the following phrase: "100 feet of shale, 20 feet of clay." The objection? "Feet of clay: cliché."

—WRL

[Note to readers, for purposes of demonstration I did not change the author's punctuation in the following paragraph.]

You used "there is" or "there are" in total about seven times in a recent column. I was taught to avoid use of "there". (I usually put the period outside the quotes). So, where you say "There are, of course, a few rules that apply to all scientific papers", I would write, "Of course, a few rules apply to all scientific papers". My statement saves space and is straight forward. It seems to me "there" denotes a location of something.

Allen V. Barker

First, American (not just ASA–CSSA–SSSA) style calls for end-of-sentence punctuation to be inside quotation marks except for a sentence like: What would your reaction be to a sentence like, "I think your research is fascinating"? All rules have exceptions. For instance, ASA–CSSA–SSSA style requires the single quotation marks around the name of a cultivar at first mention to be *inside* all punctuation.

As for the main point, if an author's goal is to write with as few words as possible, Dr. Barker would be correct. But there are other considerations in standard English—and both versions of the quoted sentence *are* in standard English, mine slightly more formal than his.

Cadence is a particular consideration that applies to written as well as spoken words. If, for example, the material that immediately precedes a sentence was a bit heavy, it is sometimes good to give readers a break by inserting words that are not technically necessary, such as, "So, then, what does all of this mean?"

Study the sentence structure used by authors you particularly enjoy reading (scientific writing, fiction, poetry, anything). Note the way individual sentences flow within themselves and together with those on either side and you will see what I mean by cadence. Cadence can be altered by adding or deleting words and also by the choice of words themselves. Poets take cadence to the ultimate step and actually use it to highlight meaning. Consider this pair of sentences by Robert Frost:

> The old dog barks backward without getting up.
> I can remember when he was a pup.

The first sentence cannot be read fast, and helps portray the age-enforced slowness of the old dog. The second sentence flows smoothly and quickly, and helps one visualize the same dog as a frisky puppy.

After stating all the above, I must add that since I read Dr. Barker's comment I now begin sentences with *there is* and *there are* much less often than I had in the past.

Dr. Barker continues:

In that sentence, you say "... a few rules THAT apply...," but later on you write "... the most common complaint I HEAR ...." I wonder why you use *that* sometimes and not others. It seems to me that it should be used every time. It does not take up as much space as *there is*.

I'm reminded of some beer ads: "Give me a light. No, a Bud light."

The correlation: "Let's talk about that. No, not *that* that, *that* that!"

Both of the examples cited by Dr. Barker refer to the word *that*, but the examples are not the same. I must write like a grammar teacher for one sentence. The first *that* restricts the sentence meaning, the second introduces a subordinate clause.

Translation: In order for a person to eliminate the word *that* from the first, he or she would need to rewrite the sentence to something like: "The stated rules always apply in this particular situation." Scientific writing requires sentences of this sort, so the word *that* needs to be used often.

The other form of *that* is not essential to the meaning. "I think we should study it," means the same as "I think that we should study it."

Most American newspapers eliminate the word *that* from sentences of this sort. I came from a newspaper background and this column is basically a journalistic effort, so that is the style I tend to use here.

I learned this style in my first journalism class when I worked as an intern on the *Ames Daily Tribune*, the commercial newspaper in the city where I went to school. The paper's editor contended the word *that* could be eliminated from every sentence without being missed. I found to my surprise that she was correct far more often than I would have guessed.

She was not correct, though, at two different sets of times. One was with sentences that fell into the first category mentioned above, when the *that* was an essential restricter. Two examples would be, "The editor of *that* paper" and "with sentences *that* fell into the first category." The other time was when the word was not essential, but still helped make a piece more readable. Sometimes this would involve cadence, as discussed above, and at other times the word can be used to clearly define the links among key words. The phrase, "I found to my surprise *that* she was correct" fits this category. The word could be crudely dropped from the phrase as presented and easily dropped if the phrase was reworded: "to my surprise I found she was correct."

Just because the word *can* be eliminated from a particular construction does not mean it *must* or even *should* be eliminated. Some sentences read better with the otherwise nonessential *that*. How can a writer decide? I use judgement. I eliminate it if the word truly adds nothing to the meaning or understanding of a sentence, and keep the word if it gives readers a clue to help make their way through the sentence. The more complicated the prose, the more cues of this sort a reader will need. If you find yourself using the word *that* a great deal, maybe your prose is more complicated than necessary.

* * *

I like to push the envelope in my writing; to try to do things people say can't be done. An analogy might be a person who is used to juggling three balls at a time occasionally trying to juggle five or seven. Perhaps you noticed that I tried that in the previous piece—a six-word sentence with four of those words being *that*. I gave myself a mental high-five after writing it and still smile when I think about it. Little tricks like that, mental games you play with yourself, help to keep writing fun and make it less of a chore.

The next article is a little along the same lines. In response to several people telling me a particular item should never be used, I explained a few times when it could. From April 1998:

Finding a place for and/or

I don't try to be contrary, but sometimes find myself rowing against the tide of opinion. One of those regards the acceptability of the phrase *and/or*. I can neither count the number of people who have told me, nor the number of places where I have read, that the phrase is always unacceptable.

Here's the way one scientist put it: "it seems to me that the term *and/or* has been designed specifically to be ambiguous. It can mean 'and' or 'or' or 'both and and or.' Those scientists who study logic must be appalled by any attempt at interpretation."

Many people agree with this scientist. They say to the writer, "Make up your mind, do you mean *and* or do you mean *or*?" The clear implication is, "You can't mean both."

Why can't you sometimes mean both? Consider this sentence: "Adding nitrogen to alfalfa may lower yield or quality, or both yield and quality of the stand by stimulating growth of grasses and weeds." The authors clearly worked hard to avoid using *and/or*, and succeeded, but ended up with several more words than necessary. Please tell me what would be wrong with, "Adding nitrogen to alfalfa may lower yield and/or quality by stimulating growth of grasses and weeds."

Contrary to the people who say "make up your mind," every now and then a person will have the need to write "one or the other or both." When such a situation exists, I believe it is acceptable to occasionally use the *and/or* phrase.

Those instances are rare, though. The same authors who avoided an acceptable use of *and/or* in the above sentence used it incorrectly in another: "Treat if you find three adults and/or nymphs per sweep." The authors meant, "treat if you find three nymphs, three adults, one adult and two nymphs, two adults and one nymph, or any other combination that totals three." The sentence does not mean that, though; because combinations of adults and nymphs totaling three are not allowed by the structure of their sentence. The sentence these authors wrote means "treat if you find three adults, three nymphs, or three adults and three nymphs" [a total of six].

To get what the authors wanted, they should have written: "Treat if you find any combination of nymphs and adults totaling three per sweep."

So, if you think you are in one of those rare situations where the use of *and/or* would be appropriate, first make certain the usage would be logically correct, and then make certain there is no better way to present the information. Let's say you are writing instructions on how to operate something and you need to insert a safety precaution along the lines of, "Whether you decide to do it this way, or that way, or a little of each, you must . . . ."

That sentence is fairly straightforward and about as unambiguous as the subject matter allows. The similar sentence with an *and/or* phrase would read something like: "Whether you do it this way and/or that way, you must . . . ." That is no more ambiguous than the first, but I believe it is more awkward. The most elegant way to present that idea would be to use more than one sentence: "This can be done in two mutually nonexclusive ways. [Present those ways, probably in two sentences.] No matter which way you choose, you must remember to . . . ." Although that construction is elegant, is it better than either of the first two alternatives? I will let you decide.

I can also see a possible need for the use of *and/or* in some legal statements to shorten the following: "Whether you do it this way, or that way, or both ways, you must . . . ." Legal statements often seem to be deliberately opaque to all but attorneys, and I make no claims of being an expert in that kind of writing, but I have noticed that many legal statements have a built-in redundancy, using phrases such as: "each and every," "cease and desist," and "ordain and establish." The *and/or* might allow a similar redundancy.

In short, there is a place for *and/or* in many forms of writing, including occasionally scientific writing. In at least some of the cases the *and/or* phrase represents an acceptable solution to the question of how to present the information. Other times it can be awkward or even incorrect.

I hope you can tell from what I've written thus far that I am not a gung ho advocate of the phrase, but I believe the phrase's gung ho opponents are incorrect. There are two reasons why I am not totally enthusiastic about use of *and/or*, even in those few places where it might be correct.

First is that the phrase is usually used incorrectly. Even when it is used correctly it often is not needed. In short, it is often used by people who do not want to take the time or effort to construct a better sentence.

The second thing that bothers me about *and/or*, even when it is used appropriately, is its occasional awkwardness.

In short, if there is a better way to write the idea, then write it that better way and do not use *and/or*. If there is *not* a better way to write it, then I believe it is perfectly all right to occasionally use *and/or*. I use it in perhaps 1% of the things I write [in the last 100 things I have written (not counting this particular column) I used it maybe once]. And I use it in 5% of the things I say.

* * *

Now I want to move onto punctuation for a few pages, beginning with this column from August 1992. For those of you who have been longing for grammar jargon, you'll find it in this piece.

Hyphens: When to use and not use them

They are tiny and seem insignificant, but hyphens have caused us more grief and correspondence than any other punctuation mark I can think of.

"You are not consistent," one author wrote on a galley proof. "Sometimes you hyphenate 2 cm, sometimes you do not."

"I don't think organic matter should ever be hyphenated," wrote another author.

"Sometimes you hyphenate soil temperature, sometimes you don't. What is going on here?" wrote a third.

A letter from one person listed 25 hyphenated terms from a recent issue of a journal that he thought should not have been hyphenated and another dozen nonhyphenated terms that he thought should have been hyphenated.

Within the past 3 or 4 months I have sent out three long letters responding to such comments. Clearly, all the above is evidence of a good reason to write a column on this subject.

Compound or unit modifiers are two or more words that together modify or explain another word. In many cases the explanatory words make no sense on their own. For instance, "water-stable aggregates" are not aggregates of either water or stables, they are aggregates that are stable in water. A "two-phase system" is neither a two system nor a phase system; a "three-neck flask" is neither a three flask nor a neck flask. In each of these cases the hyphen makes it clear that both words, taken together, explain the third.

The reason why Headquarters editors are accused of inconsistency is that the words that are hyphenated when they form a unit modifier can also be used without hyphens in another grammatical construction. For example, a person can either take a *soil-temperature measurement* or measure the *soil temperature.* The first construction requires a hyphen, the second does not.

The same rule applies when one of the units of a modifier is a number. A *2-cm diam.* pipe is a pipe that measures *2 cm in diameter*. The first construction requires a hyphen (and abbreviation of the word, diameter), the second construction requires neither.

Not all unit modifiers are hyphenated. Adverbs that end in "ly" are not hyphenated. Examples of this include *recently developed* procedure or *carefully planned* experiment.

The word *early* happens to end in "ly," but is not an adverb, so needs a hyphen when it appears in a phrase such as *early-morning* sampling.

Another category of exceptions involves terms that are so familiar that there is no chance of confusion. Hyphens are not required with any of the following terms:

- stainless steel flask (or tank or whatever)
- molecular weight determination
- high molecular weight enzyme
- melting point determination
- data acquisition system
- organic matter content
- organic C content

The last two terms were added to this list in response to one of the comments quoted at the beginning of this paper.

When a chemical name is a part of a unit modifier it is not hyphenated (example, amino acid level) unless the second word is a verb or a present or past participle, such as *methyl-substituted intermediate.* All unit modifiers with a verb or past participle as the ending word are hyphenated (*air-equilibrated* samples, *rate-limiting* step, and *ion-promoted* reaction).

If one of the words of a unit modifier is a proper name the term is not hyphenated, as in *Lewis acid catalysis* or *Fourier transform technique.* Similarly, no hyphen is needed if the two words are a non-English phrase such as in situ evaluation or ad hoc committee.

Normally a hyphen is not needed if the first of the two words is a comparative or a superlative: higher temperature reactions and lowest frequency wavelengths fall into this category. Least-squares analysis and nearest-neighbor interaction are exceptions to this rule.

A hyphen is needed if the first word of a compound is an adverb, such as well, still, or ever (for example, well-known procedure, ever-present danger, or still-new equipment).

Three-term modifiers normally require hyphens: out-of-plane distance, 3-yr-old equipment, 4-mm-thick layer, and signal-to-noise ratio.

Practically any two-word combination that modifies a noun should be hyphenated: air-dried soil, high-temperature chamber, etc.

Your staff of professional editors at Headquarters may occasionally make an error—that's why we send you the galley proofs. But most of the things that might appear to be inconsistencies are due to complexities of the English language, not to our whims or errors.

For further information on use of hyphens, see pages 37 through 39 of the ASA–CSSA–SSSA *Publications Handbook and Style Manual* (1998), pages 31 through 43 of the American Chemical Society *Style Guide*, and other style books, grammar books, and large dictionaries.

* * *

For the next foray into explaining punctuation I reverted back to my subliminal seduction method of presenting grammar rules without calling them grammar rules, and was criticized by some for not presenting the appropriate grammar rules. This was published in September 1996:

Comma use and abuse

I believed my column for July was ready to submit to *Agronomy News*. Before I passed it on to the editor, though, I read it one last time to see if there was any way I could improve it. I didn't change a single one of its 1,054 words, but I *did* eliminate one comma.

Is that a substantial change? I guess it is all in how one looks at something. One could quote the fairly famous line attributed to Michelangelo. An unnamed person had visited his studio, then dropped in again a few weeks or months (depending on the version of the story) later and saw the artist still working on the same marble statue.

"You have made no progress," the visitor said.

"Quite the contrary," the sculptor replied, "I softened this cheekbone, made the lip more realistic, and improved the shape of the left eye."

"But those are mere trifles," the critic said.

"Ah, yes," the master replied. "But trifles add up to perfection, and perfection is no trifle."

Is a comma a trifle? Someone decided recently that a comma in a sentence in the Constitution of the state of Michigan allowed slavery. The sentence read, "Neither slavery nor involuntary servitude, unless for the punishment of crime, shall ever be tolerated in this state."

That sentence, including the placement of its commas, is an almost direct quotation from the 13th Amendment to the U.S. Constitution, which outlawed slavery. As I have discussed in an earlier column, our language continues to evolve. Although both of the sentences forbad slavery at the time they were written, a person who reads them today with no sense of history would interpret them as allowing slavery for convicted criminals. So there would be no confusion, Michigan legislators and citizens moved the comma. The sentence now reads, "Neither slavery, nor involuntary servitude unless for the punishment of crime, shall ever be tolerated in this state." (It is much more difficult to amend the U.S. Constitution, so the comma has not been moved in the similar sentence in its 13th Amendment.)

In the beginning was the (spoken) word

Every language (that I know of, at least) began as spoken words, which were not written down for the first time until many years or even centuries later. When people speak, they vary the amount of emphasis they place on words and vary the lengths of the pauses between words. These pauses and variations in emphasis convey at least as much meaning as the words themselves.

As a part of the development of a written language, devices were needed to show the reader where to place the emphasis and pauses so that the written word could convey the same meaning as the spoken. Italics, underlining, bold type, and similar devices show where to place emphasis. Periods, parentheses, dashes, colons, semicolons, and commas show where and how long to pause.

Theodore M. Bernstein, in *The Careful Writer: A modern guide to English usage* (Atheneum, 1984), shows how the meanings of three sentences can change radically with the addition of only one or two commas:

a. Was that the latest dope?
b. Was that the latest, dope?
a. The Democrats, say the Republicans, are sure to win the next election.
b. The Democrats say the Republicans are sure to win the next election.
a. Do not break your bread or roll in your soup.
b. Do not break your bread, or roll in your soup.

Used properly, a comma can clear up an unclear sentence (and even help make sense of Pennsylvania Dutch speech patterns): I always believed the sentence, "Throw momma from the train a kiss" was spoken by a child from a train platform to the mother on the train. Therefore, I always punctuated it this way, "Throw, momma from the train, a kiss." When I placed that as an example in this column, though, several reviewers corrected me. The sentence is spoken by the mother on the platform to her child in the train, and should be punctuated this way, "Throw momma, from the train, a kiss."

Used improperly, a comma can inject untold confusion: "There goes the head of the department, a scholar and a dog." Is that one despicable person, or two people and one dog? Without a comma after *scholar*, it could be either.

Use commas sparingly

Bernstein in the previously cited book writes, "The tendency these days is to use a minimum number of commas. And if a writer feels the need to use a multitude of commas in a sentence, it is likely that the sentence is confused and needs recasting." I composed the following run-on sentence as an example:

> Too many commas can be as bad as too few, there are writers who string word after word with nary a period, believing that commas will make the meaning clear to all who read, they do not, because commas are what the British call a "partial stop," as opposed to a "full stop," the British term for period, and I had better bring this terrible example to its own full stop, lest you think I don't know the difference between a comma and a period.

Even that terribly written run-on "sentence" is more clear with commas than if there had been no punctuation in it at all. Clearly, as Michelangelo might have said, the comma is no trifle.

There are, of course, many specific uses for commas. So far I've shown commas used to separate phrases, set off quotations, and clarify sentence meaning.

Misuses of commas

The most common misuse of commas is with restrictive phrases and clauses. Commas should be used with nonrestrictive phrases and clauses, but not with restrictive ones. What is the difference?

A restrictive term is essential to the meaning of the sentence. "Nobel Peace Prize recipient Norman Borlaug is both a wheat breeder and dynamic speaker." Commas are not used around Dr. Borlaug's name in this sentence because his name is essential to the meaning of this sentence. He is not the only person who has received that prize, and that sentence would not be correct without his name: "Nobel Peace Prize recipients are both wheat breeders and dynamic speakers."

The following seemingly identical sentence requires commas: President Clinton's wife, Hillary Rodham Clinton, is controversial. Unless the President were a polygamist, this sentence would be just as accurate without her name: "President Clinton's wife is controversial." In this case the name of the First Lady is a nonrestrictive phrase, which means that although it presents pertinent information, it is not essential to the meaning of the sentence.

A nonrestrictive phrase in a sentence about Dr. Borlaug might be: "Norman Borlaug, Nobel Peace Prize recipient, is both a wheat breeder and dynamic speaker." The sentence without the nonrestrictive phrase would make every bit as much sense: "Norman Borlaug is both a wheat breeder and a dynamic speaker."

What you need to know

I could quote grammar books, dictionaries, and style manuals as to specific places where commas should be used, and where other sorts of punctuation are more appropriate. All you really need to remember is that punctuation is used to produce pauses. A period calls for the longest pause. A slightly briefer pause is called for by a dash or set of parentheses. Briefer than those are the pauses called for by colons and semicolons. The comma calls for the briefest pause of all.

There are a few exceptions. Sometimes series are presented in such a way that both commas and semicolons must be used: This year's officers are C. Jerry Nelson, Missouri, ASA President; A Bruce Maunder,

Texas, CSSA President; and H.H. Cheng, Minnesota, SSSA President. Even in this series, especially if you read it aloud, you can see that the semicolons call for a longer pause than do the commas.

No editor would complain if you studied the rules in dictionaries, style manuals, and grammar and usage guides and learned how to punctuate perfectly. But if you remember the basic concepts presented in this paper you know in most cases whether and where to use commas. You will be able to write with enough clarity that people will be able to understand you. In the few cases where this will not be precise enough for publication, the professional editing staff employed by the publisher can help you with the finishing touches.

* * *

From April 1994, here are the rules regarding use of another sort of punctuation marks:

Use and misuse of quotation marks

Some of you may have noticed in my December 1993 column (page 26 in this book) that I used the phrase, "what I jokingly call my first draft." From then on, I referred to my "first draft" with quotation marks around it. Use of quotation marks in this way is a device authors use to tell readers, "I'm using the word here with a different-from-normal meaning."

Think of that the next time you write an acknowledgment. Which of the following two sentences would you use?

A. I thank Jonathan Glotzgahagen for his excellent assistance.

B. I thank Jonathan Glotzgahagen for his "excellent" assistance.

Sentence B is a great way to insult Glotzgahagen or anyone else who doesn't realize that quotation marks in this type of construction change (in this case, probably reverse) the meaning of a word. On the down side, use of a construction such as that in Sentence B is also an excellent way of showing the whole world that the author does not know how to use quotation marks.

In a similar vein, if you write Jennifer Glotzgahagen had a "new" idea, you are indicating that her idea was far from new.

When I see an advertisement with a sentence like: Made with "real fruit juice," or made with "real" fruit juice, I first want to chuckle at the advertising copy writer who didn't know how to use the language. Then I wonder, instead, if I should be wary of the product; it is always possible that the copy writer knew exactly what was meant, and deliberately used a wording that would fool the public and sell an inferior product for a high price (a little like the company that advertises "genuine faux pearls").

In other words, if you want to reverse or deny the meaning of a word—as I did in the first paragraph with "first draft"—use this construction. Otherwise, don't use quotation marks around a word or phrase unless you know what you are doing and mean it.

There are several other ways quotation marks can be misused. Before I discuss any of them, I'll explain the acceptable uses of quotation marks.

Proper use of quotation marks

The most common and always appropriate way to use quotation marks is to mark off a direct quotation: Jonathan said, "Thank you for your help." To repeat what I said yesterday, "My answer is no."

Such a direct quotation does not have to be of an entire sentence: Jennifer said she would be "pleased" to help with the project. The research leader said he was "stunned" when he heard the news that he had been awarded this year's Nobel Prize. The department head said she was "surprised and shocked" to learn a scientist in her department had been accused of scientific fraud.

A usually acceptable way of using quotation marks is around a trade or product name. Such names must be set off in some way to let the reader know it is not a common name. The usual way is to capitalize it, but another is to put it in quotes: "Styrofoam" mastic. The problem with quotes here is that some people may misinterpret it as a bogus product, as discussed at the beginning of this paper. I, therefore, prefer to use Styrofoam brand mastic. The ASA–CSSA–SSSA journals, though, use simply Styrofoam mastic, allowing the capitalization alone to show the distinction. (I used mastic in this example to remind you that there are more things than just foamed styrene plastic items that carry the Styrofoam trademark. See Chapter 5 for more on use of trade names.)

A third acceptable use of quotation marks is to set off titles of book chapters, speeches, papers, reports, studies, etc.: The chapter titled, "Manuscript Preparation" in the book *Publications Handbook and Style Manual.*

A fourth acceptable use is to set off a nickname: George Herman "Babe" Ruth.

Other misuses

In spite of what you might read in the *GPO Style Manual*, most writing experts agree that one of the more common *misuses* of quotation marks is a construction that indicates, "I don't want to use such a hackneyed word or phrase as this, but I cannot think of anything better, so I'll put quotation marks around it to show that it isn't something I would normally use myself." Two examples:

- According to "conventional wisdom," this ought to work.
- The "man on the street" says it should be done this way.

This construction shows that the author (i) does not know how to use the language, and (ii) has a limited vocabulary.

A final misuse is throwing in quotation marks for no reason whatsoever: There is no "intrinsic" reason why this should not work. I've asked authors of sentences like this why they inserted quotation marks. Their reply, "I don't know, it just seemed like the right thing to do."

[I thank Pamm Kasper, formerly a Headquarters associate production editor, for suggesting this topic.]

* * *

Moving onto another item of punctuation (and at the same time having fun with the title by alluding to a line from the film *Wizard of Oz*) I wrote this one, which was published in October 1998:

Hyphens and dashes and stuff (oh my)

Most people have a reasonably good feel for how to use the standard punctuation marks; the periods, question marks, quotation marks, and so forth. (If you are wondering why I did not include the exclamation mark in the previous list you must have not read my column last month [p. 150 in this book]. The best writers hardly ever use exclamation marks, except in direct quotes.)

Another group of punctuation devices *do* cause problems, however. Those are the marks that set off a what most of us would call a *parenthetical statement*. (Fowler calls the statement itself a *parenthesis*, and the punctuation around it *parentheses punctuation*. Most everyone but Fowler calls that punctuation *brackets, parentheses, hyphens*, and various sizes of *dashes*. (What we call *parentheses*, Fowler calls *round brackets*.)

I dealt with hyphens in one of the first of the columns in this series (published in August 1992), page 138 in this chapter, so will devote most of this column to the rest of these marks.

To quote Fowler on their use: "Square brackets are the most disconnective; their main use is for an explanatory interpolation in a quotation. Of the other three, commas are suitable for [the parenthetical statement] that least interrupts the run of a sentence, and dashes and round brackets that do so progressively more."

The following four sentences use those marks progressively to illustrate this function:

- When you want to make a minor aside, just a little interruption, use commas.
- If you need to interrupt a bit more—say to emphasize a key point—then use a long dash.
- The long dash (I'll define the term and write more about it in the next section) does not give as much of a break as does the parentheses.
- I reserve use of the square bracket [did you notice that unlike when you type it, when it is printed there is no space on either side of the long dash] for major interruptions; usually author's notes or asides.

Long dash

The long dash is sometimes called *the mark of ignorance*, because most people use it when they do not know what else to use.

Editors and printers call it the *em* dash, because in an earlier age those marks were the width of a capital letter M. (There is also an *en* dash, which I will get to later, that was at one time the width of a capital N.)

Perhaps the main distinction with the long dash is that it denotes a sort of a "wait, there is more to say about that." A comma does that to some extent, but not as well. Indeed, many people use a dash when a comma would be more appropriate. Here's an example of that: "The presentations—and especially the one by Dr. Blather—impressed the audience." There is really no need for a long dash in that sentence, a set of commas would have accomplished the same purpose.

On the other hand, here is a sentence from Barry Commoner that correctly uses the long dash: "When in 1960 the stockpile was sold off—indeed, dumped as surplus—natural-rubber sales were hard hit."

You can also use the long dash in a number of ways besides

separating parenthetical material. The following examples all come from the punctuation guide in *Webster's Ninth New Collegiate Dictionary*:

- To introduce a summary statement that follows a series of words or phrases: "Oil, steel, and wheat—these are the sinews of industrialization."
- To attribute a quotation: "My foot is on my native heath . . . —Sir Walter Scott."
- May be used with a question: "Your question—it was *your* question, wasn't it, Mr. Jones?—just can't be answered."
- The dash removes the need for a comma if it falls where a comma would normally be used to separate two clauses: "If we don't succeed—and the critics say we won't—then the whole project is in jeopardy."

The usual way to designate an *em* dash in typewritten text is with a double hyphen. In other words, the previous sentence would be written, "If we don't succeed - - and the critics . . . ."

Short dash

The short or *en* dash seems to be used almost exclusively in writings about science or mathematics. Sometimes they are confused with hyphens, because many writers are not familiar with the difference between them.

When an author of a paper to be published by ASA–CSSA–SSSA uses a hyphen to denote a range of numbers, a professional editor in Madison will change it to either a short dash (if those numbers are inside a parentheses) or to a word, usually the word *to*. In other words, "Yield improvements ranged from 15 to 25%." Compare that with, "Yield improvements were impressive (15–25%)."

The short dash is not used only for number ranges, though, as the abbreviation in the first line of the previous paragraph demonstrates. That example is a slight extension of the precise rule, though, which calls for using the short dash to join nouns of equal importance. An example in law might be the Taft–Hartley Act. In soil science, Webster–Nicollet soil complex.

The short dash is also used in a number of ways in chemistry, one of which is to join two chemical compounds, as $HCl–H_2SO_4$ and also within a compound designation that contains a subscript, such as, $NO_3–N$.

One of the more complicated questions a member ever asked me regarding the use of short dashes in a number range concerned a sentence that would have required both a hyphen and an en dash. Here's the situation: In instances where a number and a unit are used together as a compound modifier, a hyphen is used, as in "a 7.5-cm depth." The question went something like this: "Suppose you have a range of numbers instead of a single number as your compound modifier and those numbers occur within parentheses?"

The matter is much simpler when the compound-modifier range is presented outside a parentheses. The use of the word *to* eliminates any ambiguity: 15- to 20-cm depth.

If the sentence were constructed in such a way that the range would fall within a parentheses, following our style would call for use of the short dash (15–20-cm depth). I likely would consider this a good enough reason to employ the word "to" there as well, following my dictum, "Always obey the style manual unless you have a good reason not to."

With more complicated numbers I would almost certainly use the word *to* whether the range appeared inside or outside of parentheses and would probably also repeat the unit: "Soil samples were collected from 0- to 7.5-cm and 7.5- to 15-cm depths."

You don't need to worry about designating *en* dashes in your manuscripts. The professional editors in Madison change the hyphens to *en* dashes as a normal part of their getting a manuscript ready for typesetting.

Parentheses

Parentheses (or round brackets) can be used within sentence, as was done here. They can also be used around one or more free-standing sentences, as I did in the first paragraph of this column. I typically use parentheses in two situations. First, as was done in the first sentence of this paragraph, is to set off something akin to an informal definition. The second is as used in the first paragraph of this column, when I used them to set off a minor aside; closely related to the main topic, but enough different to require separate treatment.

Square brackets

I reserve square brackets for asides to text that are a more major break than called for by parentheses. The distinction is easier to recognize than to write about. As a rule of thumb, though, I would say that if in speaking I would preface the phrase with "by the way," I would probably separate that phrase from the rest of the text with square brackets.

Of course, both parentheses and square brackets also have specific uses in mathematics that are separate from what are written about in this column. In one regard, the usage is the direct opposite. When a series of them are used in math, the square brackets are outside the parentheses: $[5 + (3 \times 2)]$. In text, the order is the opposite: " . . . (as Stelly used to say [quoting J. Fielding Reid] . . . .)"

Conclusion

Perhaps the main thing to remember about all these forms of punctuation is that there is not always a clear distinction between them and experts differ on which usage is correct. For example, the sentence with the long dash, above, that I gave as not being a correct usage was a direct quote from an authoritative dictionary. Many of the square brackets I used in sentences in these columns when they were first published by ASA–CSSA–SSSA were changed by the newsletter's editor to standard parentheses. Do the best you can, the editors who work on your paper will do the best they can, and hopefully the resulting product will be not bad.

* * *

The next column elicited a few objections after it was published in September 1998, but I stand by every word in it.

Why you shouldn't use exclamation marks

The best writers hardly ever use an exclamation mark, except in a direct quotation.

An exclamation mark tells your readers, "By the way, that sentence you just read was supposed to have been forceful." If you need to tell your readers that, there is something drastically wrong either with the sentence itself or with your perception of it.

Take another look at any sentence you've written that ends with an exclamation mark. If it was forceful to begin with, it doesn't need an exclamation mark at its end. If it was not forceful but you want it to be, tacking an artificial mark on its end will not make it so. The only way to meet your goal of making it forceful would be to rewrite the sentence.

Anytime a writer feels the need to tack an exclamation mark on the end of a sentence, he or she is admitting a belief that the sentence was not written very well.

First way to misuse the marks

There are two ways a writer can misuse an exclamation mark. The first is to use the mark in an attempt to add strength to a poorly written weak sentence. This use of an exclamation mark is roughly equivalent in speech to saying, "I was unable to convince you the first time I said it, so I'll say the same thing again more loudly."

Of course, that doesn't work. What you need to do instead of saying it again more loudly, is to say it the first time more clearly, or more convincingly, or with more authority.

Let's take an example: "Their hypothesis was not supported!"

The first thing I ask myself when I see a sentence like that is, "I wonder why the authors put an exclamation mark there? What are they trying to tell me with that mark that they couldn't with their words?"

Let's rewrite the hypothetical statement in a way that would likely not prompt such a question: "Hundreds of scientists in dozens of labs around the world tried to duplicate their results under every possible combination of environmental conditions. Not a single one of those efforts produced the same results as were claimed in their paper."

No exclamation mark needed there.

Second way to misuse marks

The second way to misuse exclamation marks is to tack them onto the end of statements that are already both well-written and forceful.

Let's say the author wanted to conclude the "Hundreds of scientists . . . ." sentences with the following summary statement: "In other words, their hypothesis was not supported." In this context this formerly weak sentence has become an excellent summary statement. But an exclamation mark will not add any emphasis to the sentence beyond the considerable amount of emphasis it already contains.

You often see scientists make this sort of error when they are writing for a nonscientific audience. A soil scientist might begin a farm magazine article with a sentence like, "If you're not aware of the soil type where you site a feedlot, you could pay for your mistake for years to come!" Or, an entomologist might begin a newspaper column with, "Watch out for mosquitoes!" Exclamation marks detract from these and similar sentences.

Where you see them

I rarely see exclamation marks in traditional scientific papers in journals or scholarly book chapters. I *do* see them, as mentioned above, in outreach (extension or whatever) literature written by scientists.

I also see them by the ton in quasi-scientific papers written by nonscientists. For instance, exclamation marks abound in literature written by animal rightists, out-of-mainstream environmental activists, and persons of that sort. These persons often don't have much evidence or documentation to back up their statements, so instead they attempt to use gimmicks to try to get their opinion across as convincingly as they can. It is as though they are saying, "If I can't convince you with my facts or with the logic of my presentation, maybe I can convince you with the decibel level at which I make my presentation."

How to add emphasis

I've written a number of times of the need to simplify one's prose and not use needless words. I stand by all those statements, but this column concerns necessary, not needless words. An exclamation mark will never substitute for a word or group of words that are essential to the meaning of a thought or sentence.

If you want to present something forcefully, choose powerful words. Generally, these will be the smallest of any group of synonyms. For instance use, "I will go," instead of "The person who is now speaking resolves to egress." After you have chosen the best words for what you want to say, group them into the most forceful sentences you can draft. If you do that, you will never need to tack on an explanation mark.

You may have noticed one device I *did* use in this column, though. That was the italicization of one or more words. I italicized a word in the first sentence of this paragraph, and also the word *do* under the subhead "Where you see them." Such a device is completely appropriate in prose to help the reader fully understand the meaning an author is trying to convey in a written sentence. Another example of this sort might be as follows, "Well, he *said* he was the one who had made that discovery." The writer's skepticism is clearly conveyed by the italics; the sentence would have been ambiguous without the use of that typeface.

So be on the lookout for exclamation marks in anything you have written. For every 100 you find, you probably need no more than one. More likely, you don't even need *it*.

* * *

Some use too many exclamation marks, others use too many capital letters. This column, from October 1992, tells why you should not:

Capitals, capitals, cut the capitals

After College, where I majored in Journalism, I took a job on a Newspaper as a General Assignment Reporter. After a few months of Experience I was promoted to Education Reporter. By my Second Year there I became a Copy Editor and also served as the Business Editor . . .

What's wrong with that first paragraph? Far too many capital letters. Such extravagant capitalization is the norm for a few languages, including German, but in the American version of English the only three words that should have been capitalized were the first one in each of those sentences.

Each year I edit the biographies of Fellows and award winners for publishing in the May-June issues of three of our journals. That first paragraph was inspired by the way some of them are written. But the material in this column comes from, and applies to, far more writing than simply biographies.

Certainly you are proud of your profession, but agronomy, crop science, soil science, astronomy, medicine, etc. are not normally capitalized. No doubt you are proud of the fact you are a professor or associate professor or division leader or crop consultant, but those words are not normally capitalized either. I learned about capitalization of job titles the hard way when one of my first employers said to me, "You can start capitalizing the word *reporter* after you regularly capitalize the words *secretary* and *janitor*."

Even words like president and department head or department chair are not normally capitalized. There is one person in the nation who can be called the President. That person's address is 1600 Pennsylvania Ave., N.W., Washington, DC. Everyone else is a president, whether that be president of a university or a business or a service club.

Names and titles

"But wait a minute, Luellen," I can hear you asking. "I distinctly saw the weasel word *normally* in several of your sentences earlier in this column. Let's hear the exceptions."

Okay, Dr. H.H. Cheng is president of the American Society of Agronomy [when this was first published]. The term of President Cheng ends this fall, when he will become past-president. What I just demonstrated was that a word may be capitalized whenever it is used as a person's official title. In other words, when President Cheng becomes past president, his title will be Past-President Cheng. Similarly, Jerry Hatfield is editor-in-chief of ASA. His title is ASA Editor-in-Chief Jerry Hatfield. If Dr. Johnson is the head of a department, her title is Department Head Johnson.

Many years ago the Societies decided to desex the title *chairman*, and shortened it to Chair. This word is generally capitalized any time it might be confused with the item of furniture. I did not capitalize it three paragraphs earlier where I wrote *department chair*, because there was little or no chance of confusion there. Similarly, I capitalized *societies* at the beginning of this paragraph to make it clear I was referring to three particular professional organizations, not society in general. I capitalized Fellows in the third paragraph of this column for the same reason.

The seasons of the year are not normally capitalized. An exception to that are the words *spring* and *fall*, which also have other meanings. By all means, anytime there is a possibility of confusion, capitalize the word *spring* when you refer to that season of the year: "This spring rusted after it got wet last Spring in the water flowing from that spring." Most authors, rather than capitalize the word *fall*, will substitute the word *autumn* (lowercased) whenever there is a possibility of confusion.

Places and things

Words like *university* can be capitalized in two different situations. The first is obvious, when it is part of the official name—the University of Wisconsin. The second is when you are referring to a specific university. For instance. I did my undergraduate work at Iowa State University. While at the University I served as sports editor of the student paper. The capitalization of it in the second of those sentences makes it clear I am referring to Iowa State, and not some other university I may also have attended.

Words like agronomy and soil science may be capitalized when they are a part of an official name, such as the American Society of Agronomy or the Soil Science Society of America. Similarly, if a particular university calls its crop science department the Department of Crop Science, then those two words (and the word department) would be capitalized in that title. And, as with *university* in the previous paragraph, you can occasionally capitalize the word *department* in a sentence to make it clear you are referring to one and only specific department.

I have noticed that at least two universities also capitalize the word *the* in their titles: The Pennsylvania State University, and The Ohio State University. No doubt there is a strong historical reason for this, but I have no idea what that reason might be. Anytime I need to use their names in an official capacity I reluctantly go along with them. Anything short of that, though, I tend to either lowercase the word *the* or avoid using the official name, and instead substitute *Ohio State* and *Penn State* (with initial caps on each word).

A large number of books and glossaries are available that give the acceptable terms used in a particular field of study. Some of these acceptable terms involve grammar, scientific nomenclature, and botanical names, all capitalized as I just did. Some soil scientists say a particular list of their acceptable terms should be capitalized as follows, SOIL TAXONOMY. Certainly, whenever those two words are used as a part of a title of a book, they will likely be given initial caps (depending on how and where that title is published). But the rest of the time those two words should be written in lowercase letters. I'm sorry if the contents of this paragraph hurt the feelings of any soil scientist. To compensate, I'll make you a deal, if you stop writing soil taxonomy in all caps, I promise never again to write GRAMMAR as I just did.

* * *

Perhaps one of the most confusing aspects of English is not that there are many possible ways of spelling things, but that most of those ways are correct under one situation or another. I tried to explain that in this column from January 1998:

How there can be spelling variations

Here is an exchange of letters that will be of interest to more than the original author. His letter is reprinted with his permission, my reply is slightly expanded from the original:

Dear Mr. Luellen:

I would like some expert advice from you on the word, *germplasm*, or is that *germ plasm*?

In our company literature we have used *germplasm* so consistently that only those who get pleasure from reading the dictionary even consider challenging whether this is one or two words. Enjoying your *Agronomy News* column as much as I do and the sound advice and direction you are able to impart with each issue, I figure you'd be able to explain the difference between Webster and *Crop Science*?

All dictionary references list *germ plasm* while *Crop Science*, and those of us in Pioneer use *germplasm* (one word). It will be helpful to have a valid reason to differ from Webster, and right now I don't have one. I would appreciate your rationale used for the *Crop Science* use of *germplasm* (one word).

I look forward to your response. Thank you.

—*Clive Holland*
Pioneer Hi-Bred International
Johnston, IA

There is no omnipotent authority over all English speakers and writers, no person who commands, "You WILL do it this way!" Contrary to what many believe, dictionaries do not dictate the way something should be done, but rather document the way a particular segment of the population (its readers, and educated individuals) does something.

There are obvious differences between large population segments; British and Americans, for example. I don't mean those differences as much in this reply as I do the other differences within nations that are far more substantial than most people realize.

The compilers of general-usage dictionaries have apparently determined that most of the people they deal with and publish for write *germ plasm* as two words. So that is the way they present the word(s) in their pages. The scientists from all around the world who make up our three Societies have said they prefer to spell *germplasm* as a single word. So the CSSA *Glossary of Crop Science Terms*, published the word that way in its pages and this is the spelling used in all Societies' publications.

If you ever find yourself with lots of time on your hands and nothing else to do, study a varied group of publications specifically looking for nuances and differences in word use and spelling. Depending on your point of view you will be either fascinated or appalled by what you find. For example, until recently the *Christian Science Monitor* never used the word *died.* In that newspaper, when people came to the end of their mortal lives they did not die, they *passed on.*

I once worked on a newspaper that spelled the fall holiday, *Hallowe'en.* While I worked there, I spelled the word that way in anything I wrote for that paper, but never anywhere else while I worked there, and have not spelled it that way since, (except in this reply).

The *Chicago Tribune* used to have a unique way of spelling many words; *tho* instead of *though*, *burocrat* instead of *bureaucrat*, *thru* instead of *through*, and so forth.

The Associated Press does not double up the final *l* or the final *e* in words, so in most newspapers you find words like *eventualy* and *employe*.

You will see disagreement among scientists as well. For example, the scientific community is just about equally split on whether *liter* should be abbreviated with a lowercase or uppercase letter *L*. Your Societies, among others, believe it should be an uppercase letter so is never confused. A number such as 25 l might be ambiguous, but 25 L will never be. There is also disagreement whether the term just mentioned should be spelled *liter* or *litre*, with ASA–CSSA–SSSA members opting for the first.

Another place where you see variation is in the common names of many plants. The Societies use the singular for all common names. For certain crops, such as corn or wheat, this poses no problem at all. Some other agricultural/science publications still use the plural for a few crops, such as *soybean* and *pea*. So when I write about those crops for ASA-CSSA-SSSA I spell them as I just did, without the final *s*. When I write about them in for certain other publishers, though, I spell them with the final *s*, and make the words *soybeans* and *peas*.

The fact that ASA–CSSA–SSSA style differs from Webster (the name used here as a generic for any popular dictionary or another publisher) in the way to spell *germplasm* is no more surprising than the fact that British and American scientists differ in the way they spell some other word common to both. There are many, of course but the one I will highlight here is *fetus* and *foetus*. Which spelling is correct? That depends on whether you are writing for a British or an American journal.

While most of these spelling variations are little more than curiosities, the one just given can wreak havoc when a person searches an index of a book for that particular word. The spellings are so substantially different that they would appear some distance from each other in a typical index. Unless the person is aware of the spelling variation and checks both possible spots in the index, he or she might incorrectly conclude that the book contains nothing on that subject.

So then, which spelling of *germplasm* is correct? If you are a crop scientist writing to be read by another crop scientist, the way I just spelled it is correct. If you are writing for a publisher who adheres to the Webster way of spelling it, then the other would be.

* * *

All letters I received (and still receive) are answered privately. Many also found their way into either an entire column, like the previous, or collected, like the following column from July 1999:

Readers ask questions about articles, brackets, and tense

I was reviewing your column in the February issue [next chapter, p. 185] and had a question on the use of *a* vs. *an*. I was always taught that it was determined by the first letter of the next word. If it is a vowel, you use *an*, if it is a consonant you use *a*.

Is this rule wrong, or just no longer used?

— *Gary Custis*

I wrote: "The idea is to eliminate a doubling-up of a particular sound" to avoid grammatical jargon. In hindsight, I should have used the words *vowel* and *consonant* in this case. The rule you state is fine as far as it goes, but falls short, as I should also have mentioned, when the *a* or *an* come before an abbreviation. I don't know of a single person who would say or write, "a F test" or "a X-ray" or "a NSF grant."

I'll use the longest of those three to explain. If all you ever did with any abbreviation was write it, there would be no problem with "a NSF grant." The problem comes with speaking. If you spoke the full name you would obviously say, "a National Science Foundation grant." But if to be grammatically correct with the abbreviation, you would say "A N S F." Unless you spoke that carefully, the listener could have difficulty sorting out what you meant. Saying, "an NSF grant" helps the listener understand.

And, because most people write the way they speak, the written form, "a NSF grant," just doesn't seem right. The same is true for "a F test" and "a X-ray." Both are grammatically correct, but don't feel right because of the so-necessary convention in speech.

Technically, I do not believe there is a rule of grammar that says abbreviations are an exception to the "a vs. an" rule, but this is one of those conventions that has developed out of practicality, if not necessity.

[A subsequent letter, not published in this book, suggested another possibility; extending the vowel rule to make it a vowel-sound rule. After all, the letters F, N, and X sound as though it were written ef, en, and ex.]

Another use for brackets

I have been a fan of your column for some time now. It is THE thing I read in the monthly newsletter. I do, however, have a question for you.

You discussed the use of [ ] brackets in a recent issue (Oct. 1998 and previously in this chapter) in parenthetical statements, but didn't go into the use of them in quotations. Could you please let me know if there are rules about the use of brackets in quotations?

I use the brackets to indicate that I have changed the tense of a quotation or other grammatical change so that the quote is streamlined into the text. I have also used the brackets when I have removed words from a quotation, either because they were not of relevance or again to bring the grammar into line with the text. I have found the Manual of Style a bit thin on this topic.

— *Antoinette WinklerPrins*
University of Wisconsin-Madison

The use of brackets you ask about in quotations is entirely different from their use to set aside parenthetical statements, which is why I didn't mention it in that column. Perhaps you did not notice, though, that although this use was not specifically mentioned, it was demonstrated. I repeat here one paragraph from that column:

To quote Fowler on their use: "Square brackets are the most disconnective; their main use is for an explanatory interpolation in a quotation. Of the other three, commas are suitable for [the parenthetical statement] that least interrupts the run of a sentence, and dashes and round brackets that do so progressively more."

Everything you wrote about the use of brackets in quotations is correct; they are used to denote a partial summary, correction, or substitution. Another example might be: Then he said, "What in the name of [deleted] to you think you're doing?"

Past or present tense

I remember that you wrote an earlier column on verb tense. What is the SSSAJ policy on verb tense in scientific papers? I am editing a paper that has much of the Results section in present tense. It is awkward to read but that may be due to convention.

— *Chuck Rice*
Soil Microbiology
Kansas State Univ., Manhattan

I'm not sure whether the information I'm about to give is policy or convention or just an opinion. In any case, it is valuable advice.

Generally speaking, use past tense to describe how an experiment is set up: "Here's what we did and how we did it." That would apply specifically to the Materials and Methods section, but also to virtually any such description in any part of the paper.

Use present tense when you report your findings and implications: "This is what happens when this is done to that." If you don't, you might sound like, "We thought it was this, but now we are not so sure."

I agree with you that too much present tense can sound stilted. Therefore, it is alright to begin a sentence in past tense, and then change: "We found that this is what happens then this is done to that." Always be careful to avoid giving an incorrect impression, though. If you must choose between a slightly stilted sentence that is accurate and a more readable sentence that is ambiguous, always opt for the more accurate.

Word has two meanings

Here is a suggested column item. Believe it or not, Webster gives the definition of *biweekly* as EITHER every 2 wk or twice weekly. From now on I plan to use 14-d interval.

— *Lisa Al-Amoodi*
Assistant editor
Headquarters Office

From a reply to one of Ms. Al-Amoodi's query letters:

"The word *biweekly* understandably causes misunderstandings but I couldn't think of a clearer, yet concise, descriptor. Perhaps biweekly is inappropriate to use since it is often misused, leading to misunderstandings. Here at the university, staff are paid biweekly, that is, once every 2 weeks. As for my manuscript, the ratings were collected once every 2 weeks, or twice monthly. However, twice monthly does not accurately describe the ratings which were collected at 14-day intervals: twice monthly could mean twice in random pattern during a given series of months. The *American Heritage Dictionary* defines biweekly as once every 2 weeks, then describes twice weekly as semiweekly. Perhaps you or William Luellen could include a discussion of the use of biweekly versus semiweekly in one of the *Fine-Tuning Your Writing* columns."

— *John C. Stier*
Horticulture Dep.
University of Wisconsin - Madison

I have never liked biweekly (or semimonthly) either, and can't think of a better solution to this dilemma than that given above by Ms. Al-Amoodi: "14-d interval."

* * *

As you might expect, several people wrote and suggested use of the British term *fortnight*, which means 2-week period. This term is fine for people who know what it means. I have rarely seen it published in scientific literature or heard it spoken at scientific meetings. In short, it meets the definition of jargon, discussed in Chapter 7, which means it should be used only when writing for people who are familiar with it.

Another grammar problem involves agreement between subjects and verbs. Two types of agreement that are needed—positive-negative agreement and number agreement. This next column, first published in March 1997 and updated for this book, deals with number agreement, and prompted a strong response, as you will see:

Getting subject and verb numbers to agree

Clay A. Robinson, of West Texas A&M Univ., Canyon, asked me to write about subject-verb number agreement, especially when there is a modifying phrase.

Modifying phrases complicate a sentence. They appear virtually anywhere and can come before, between, or after subject and verb. The important thing for this discussion, though; modifying phrases sometimes change the number involved in the subject and sometimes do not.

How can you sort it out and make sure subject and verb numbers agree? Ignore the peripherals and ask yourself how many are doing whatever the sentence or part of a sentence discusses. The answer will tell you the sort of verb to use. In other words, "one person or thing *is* doing something," "two (or more) people or things *are* doing something." Thus:

- I am going.
- I am going with John. (This modifying phrase does not change the number of the subject. If it helps, you could also read the sentence as "I am the only person going with John.")
- John is going.
- John is going with me. (Same as above, no change in number.)
- John, but none of the other staff members, is going. (Still no change in number. By the way, this is a common sentence structure in scientific writing: "Vernal, but none of the other alfalfa cultivars, is . . . .")
- John and I are going. (Now two are acting so you need a plural verb. The subjects will be plural for the following five items.)

- John and 400 other people are going.
- Four hundred other people and John are going.
- John, 400 other people, and I are going.
- John and his dog are going.
- We are going.

Modifying phrases can sometimes make it difficult to find the subject, depending on the way a sentence is written. In the following sentence the order is simple and the quantity of the subject is easy to discern: "Of all animal species, *Homo sapiens* is the most intelligent." Turn the order of that sentence around, though, and that becomes more difficult to see: "*Homo sapiens*, among all animal species, is the most intelligent." It grates on some people's sensibilities to see *species* immediately followed by *is* instead of *are*, but in each of those sentences the subject is *Homo sapiens*, which is singular. The modifying phrase *all animal species*, which of course is plural, does not change the number of the subject.

Now consider the phrase from the previous sentence: ". . . in each of those sentences the subject is . . . ." *The subject*, not *sentences*, is the subject of that phrase, so it gets a singular verb.

The word *each* is generally a clue that the subject is singular. The word *every* is a less certain clue. While it often requires a plural verb, nobody with any sense of the language would write: "Every sample were contaminated." The reason for this is because more and more people today use the word *every* to mean "every *one*," which, of course, is singular. That was the case in the example just given, which was written to mean, "Every one of the samples was contaminated."

Tricky words

More difficult situations involve words that only appear to be singular. *Data* sounds like a singular word but means (to physical and biological scientists), "pieces of information" and therefore requires a plural verb: "Our data are sound." If that does not seem correct to you, mentally replace the word with its definition: "Our pieces of information are sound." In case you ever need to use them, the singular form of *data* is *datum* and the plural form of *symposium* is *symposia*. Use, "one growth *medium*," and "several growth *media*."

One sentence in the previous paragraph may have thrown some people. I'll bet at least one reader thinks I should have written, " . . . the singular form of data are datum." As I wrote earlier, you need to find the subject, which in this sentence is *the singular form*, not *data*. Turning the sentence around might make that more clear: "For *data*, the singular form is *datum*."

One more thing must be written about the word *data*. While, as mentioned above, most physical and biological scientists use *data* as a plural to mean "pieces of information," many in the humanities use the word as a singular collective to mean "a mass, body, or collection of information." You need to remember that if you write for publications in that field of study.

Collectives

[In the example I'm about to give the verb, *reached*, is the same for both singular and plural nouns—I reached, we reached, they reached—so I will highlight pronouns instead of verbs to show the differences I'm trying to point out.]

Perhaps the most confusing words are the collectives. These words collect people or things into a unit, and are therefore singular: "The *committee* reached *its* (not *their*) decision." The foregoing is true because the full meaning of the sentence is: "After the members of the committee reached their individual decisions, they voted as one."

"Wait a minute," you might ask. "Your previous sentence contained 'committee–they.' Your sentence just before said that was incorrect."

The subject of the second of those two sentences is not the singular word *committee*, but the plural *members of the committee.* A less cumbersome word order shows this more clearly: "The members of the committee voted as one after reaching their individual decisions."

Now for some added confusion, collective words can also be used in some contexts to mean all the individual members of the group. In such a case the normally singular collective noun becomes plural and requires a plural verb. Writers who know what they are doing can use the above information to create grammatically correct but silly sentences like: "The committee went through *their deliberations* before reaching *its verdict*."

Here is a practical case: "A *pair* of questions in this exam confuses me," but "*five* questions in this exam confuse me." *Pair* is a collective. *Five*, like any number larger than one, is usually not a collective.

But (and this is important) numbers larger than one can become collectives: "I have three ones and a five." The *five* is used here not as a number, but as the name for a particular denomination of currency. In short, it has became a collective.

Now, suppose the minimum amount of a reagent needed to catalyze a particular reaction is 500 mL. Like the example with money, in this context 500 mL has changed from a number to a specific unit; it becomes a collective and requires a singular verb. Thus: "In this formula, 500 mL is the minimum amount of reagent needed to catalyze the reaction."

Many would argue (and I would not entirely disagree—the language is evolving on this) that the above rule should be extended to phrases like the following: "Then 350 mL [or whatever the number] was added" (instead of were added). Those who argue for this use contend that the person did not add 350 individual units of 1 mL each, but added the entire 350 mL as one unit. As long as you are consistent, most editors will go along with your choice of either singular or plural verbs in this situation.

Of course, you could avoid the problem entirely if you wrote in active instead of passive voice: "Then we added 350 mL . . . ." "Next you add 350 mL . . . ." "Add 350 mL . . . ."

Finally English contains words, such as *company*, that can be either a plural (like *data*) or a singular collective, depending on the context: "I had company for dinner. They ate me out of house and home." "She owns a company. It makes computer software."

Final exam

Let's conclude this column with a one-question quiz that isn't as simple as it looks. Which is the correct verb in the following sentence? "A total of 49 papers (was/were) presented at the recent conference."

What was/were presented? Forty-nine papers. The phrase "a total of" does not change this number, so the correct verb would be *were*.

I can almost predict that someone will counter, "But doesn't 'a total of' by definition sum up individual parts and make it into a unit, which would therefore require a singular verb?"

Under some circumstances, yes, but consider the location of that phrase in this sentence. The author may have merely wanted to place a meaningless word or phrase at the beginning to avoid writing out the number, "forty-nine." We see phrases like these all the time:

- "Approximately 49 papers were presented . . . ."
- "Next, 49 papers were presented . . . ."
- "After that, 49 papers were presented . . . ."
- "During the discussion, 49 papers were presented . . . ."
- "Then 49 papers were presented . . . ."

Because none of the above modifying words or phrases treat the number as a unit, I do not believe in this context that "A total of . . ." would either.

Writing in active voice would eliminate this problem, too: "Members presented 49 papers at the recent conference." "Scientists presented 49 papers at the recent conference."

So the answer to Dr. Robinson's initial question is that some modifying phrases change the number in the subject, but many do not. The key is to ignore everything else and ask yourself how many are doing whatever the sentence discusses. Keep in mind as you consider the subject that some words can be tricky in regard to the number they represent. Once you determine the number involved in the subject, then choose the appropriate verb to match that number.

* * *

As mentioned in the introduction, I thought that previous piece, had presented the final word on the subject in that little final exam, but dozens of people disagreed, some of them vehemently. So 3 months later, June 1997, I wrote another column containing some of those letters. This is the column I was referring to at the beginning of this chapter when I wrote that grammar jargon got in the way of communication. As you peruse the following delightful exchange of letters, I'll let you make up your own mind about whether the letter-writers or I am correct. But please send no more letters (unless you agree with me).

I told you the quiz was not simple

My March 1997 column concluded with the following one-question quiz, which I warned was not as simple at it looked:

Question: Which is the correct verb in the following sentence? "A total of 49 papers (was/were) presented at the recent conference."

The answer: What was/were presented? Forty-nine papers. The phrase "a total of" does not change this number, so the correct verb would be *were*. [More of the original answer will be repeated below.]

Several readers wrote to disagree. On the assumption that more people disagreed than wrote, I present (anonymously) the main objections to that answer and my replies to their objections.

Objection

Of course, the way you rearranged the sentence it should have a plural verb, but this does not treat the sentence the way the author intended it. *Forty-nine* is a red herring, the subject of the sentence is *total.*

My reply

Are you certain that was the way the author intended it? I don't think so. My reason for thinking this is this question, Which was presented at that conference, *total* or *papers*? Consider the sentence without the opening phrase: "Forty-nine papers were presented at the recent conference." I do not think there can be any dispute that *papers* is the subject of this sentence and *were presented* is the verb. The words *forty-nine* are part of the subject, an adjective modifying the noun.

The sentence as given in the previous paragraph is complete within itself, so I do not believe the author intended to have the addition of a meaningless phrase to its beginning change the meaning so radically that one of the words in that phrase becomes the new subject. I will explain this stance in more detail as I deal with the next objection.

Objection

Isn't "of 49 papers" a prepositional phrase? If so, what happened to the rule that objects of prepositions cannot be subjects of sentences?

My reply

Yes, of course it is. But I believe it is a prepositional phrase that the author inadvertently created and did not realize he used. The author wanted to place a meaningless word or phrase at the beginning of the sentence so he would be able to write the number as numerals instead of words. As I stated in the original answer, we see such phrases all the time:

- "Approximately 49 papers were presented . . . ."
- "Next, 49 papers were presented . . . ."
- "During the discussion, 49 papers were presented . . . ."

I believe the author thought "a total of" would work the same way. Therefore I ignored the author's inadvertently created phrase and instead concentrated on the author's intention when I wrote as a part of the original answer, "Because none of the above modifying words or phrases treat the number as a unit, I do not believe *in this context* that 'A total of . . .' would either" (emphasis added). I'll write more about this after dealing with the next objection.

Objection

Why all of a sudden are you advocating the use of a meaningless word or phrase in a sentence? It seems to me that virtually every one of your columns before this one extolled the virtue of simplicity in writing and the need to eliminate all meaningless words and phrases.

My reply

There is a vast difference between *meaningless* and *unnecessary*. I explained several uses for meaningless-but-necessary words and phrases on page 114–116 and 180–182. The meaningless words and phrases given above perform the necessary task of allowing the number that follows to be written as a numeral. In addition to being easier to comprehend, a numeral usually shortens the sentence, as the following two paired examples illustrate:

- Nineteen-ninety-seven marks the 25th anniversary . . .
- The year 1997 marks the 25th anniversary . . .
- Forty-nine papers were presented . . .
- Some 49 papers were presented . . . .

Final comments

Was I wrong to concentrate on the author's meaning instead of his poor choice of words? Strictly from a grammatical point of view, yes. And, in fact, I corrected it when I came across that actual sentence in a paper I was editing.

I retained it incorrectly in the quiz (warning everyone at the outset that it was not as simple as it appeared) because I wanted to make the point about the duty of the reader to try to find the meaning behind an author's words. In recent years I have heard and read far too many commentators ignore a person's ideas and instead attack that person's poor choice of words.

These attackers remind me of the person who did not see a forest because too many trees were in the way. Particularly in this day and age, when instruction in writing and rhetoric is so poor in so many elementary and secondary schools, attacks like these always strike me as an intellectual cheap shot.

The underlying purpose of any communication is to take an idea from one person's brain and transfer it, as intact as possible, into another person's brain. Part of the task facing the person with the idea is to select the best communication method to use in a particular situation. That best method may not involve words:

- The best assembly instruction to give with a jigsaw puzzle is a picture of the assembled puzzle.

- If you are driving any vehicle larger than a bicycle, the best way to warn a person standing in your path of your approach is to honk your vehicle's horn and maybe also flash its lights.

Most complex ideas are written either to be read directly by the person you want to inform, or to serve as a guide for another presentation form, such a script for an audio or TV broadcast or instructional film.

A double responsibility faces any two people who try to communicate with words. The writer has the prime responsibility to transfer his or her ideas as intact as possible into words, word combinations, illustrations, tables, etc. Suppose the writer falls short, though? The reader has a responsibility too; that is to discover the writer's ideas as he or she reads or listens. Written communication involves a minimum of four aspects:

- What the writer means
- What the writer writes
- What the reader reads (or listener hears)
- How the reader (listener) interprets it.

The only way to correctly complete the fourth step is with the desire to discover the first. When communication falls short, the primary responsibility rests with the writer, there is no disagreement about that. But just as it is the writer's responsibility to do the best possible job of presenting an idea, it is the reader's responsibility to do the best possible job of discovering that idea.

Reviewers, editors, etc., who come between a writer and a reader, share a little of both responsibilities, which compounds these four aspects.

* * *

For more on this subject, see the article beginning on page 93.

Chapter 7

Words

Words. Where would a writer be without them? Used properly, their meaning is precise and communication is not only possible, but fruitful. Used elegantly, spirits can soar and information can be transferred. Used incorrectly, communication is disrupted at best and sometimes prevented.

Mistranslation of a single word caused a serious disruption in relations between the USA and the USSR during the Cold War. During a summit conference, President Kennedy told Premier Khrushchev that they needed to be careful, lest a miscalculation on either side lead to serious consequences. Unfortunately, there is no true synonym for the word *miscalculation* in the Russian language. The translator used a word that literally meant "you don't know how to calculate," and in the vernacular was used as a pejorative roughly equivalent to "you are stupid!"

Rather than accidentally misuse the language, too many politicians today deliberately misuse and cheapen the language to the point where there was even a question at one time concerning the meaning of the word *is*. Others with other agendas have also wreaked havoc in one way or another.

Too many who are less than professional in their speaking and writing take words too much for granted, and through stupid errors have severely cheapened our language. Probably more of my columns and other writing have been devoted to words than to any other subject. Even with all that, I have not begun to exhaust the ramifications of the subject. How does a person who wants to use words properly cope with all the above? I believe he or she must do two things. First is to become aware of the precise meaning of words and use them only in that way. Second is to become aware of the way other people have changed some of these meanings and to decide how to cope with that change.

I begin this chapter with a discussion of the ongoing evolution of the meaning and usage of words in English—particularly of the American version of English. This column provoked quite a bit of interest when it was first published in February 1996, and may be news to some people still today:

Knowing when to hold and when to fold

You don't need to go all the way back to Shakespeare to find evidence that English is evolving. An 1895 manual contained this phrase, "Go to a workman and cause for him to make for you . . . ." Today that phrase would probably have been written, "Have a vendor make . . . ." Each was acceptable in its day; but who knows how that phrase will be written in another hundred years; will it be something even more wordy than the 1895 version, or something simpler, such as "Get . . . ."?

With so few people knowing how to write well, but so many people writing and speaking nevertheless, changes in English are taking place around us all the time. Not all of these changes are for the better; far too many words are so misused that they have become meaningless.

Ecology has lost its own meaning and become synonymous with *environment*: "We all need to do our part to protect the ecology."

Propaganda, which had meant any attempt by anybody to convince anyone of anything, has come to mean *lies by big business, big government, or anything big and bad.*

Rhetoric, the art of speaking well, has come to mean *empty blather*.

Word usage, too

Not only has the *meaning* of many words been rendered useless, so has the *application* of other words. For example, what sort of miracle took place that gave inanimate objects the ability to grow? We hear the President say he is "trying to grow the economy," and that an entrepreneur has set out to "grow her business."

That misapplication of *grow* reminds me of a word misuse by one of my sons when he was about five years old. He said of something he didn't like, "I'm going to extinct it."

A person can grow a flower, but the correct way to phrase the statements mentioned above would be the President is "trying to make the economy grow" or the entrepreneur has set out to "make her business grow." This difference may seem subtle or even silly to some, but it is important to anyone who cherishes the language.

What can we do?

What can a person who tries to communicate clearly do about word and word-meaning changes and misuses? There are three possible responses; two, in my opinion, worthless. Let's mention the worthless ones first.

- Go along with the crowd: "I know this isn't the correct use of the word *grow* [or *parameter* or whatever], but everybody uses it this way now so I might as well too."
- Fight for the old way or the old meaning: "I don't care what everyone else says, the word *gay* means happy, and, by golly I'm going to use it to mean happy until the rest of the world comes back to the correct usage again."

Choose your battles

If you don't care to either (i) give up and go along with a crowd that is doing something stupid, or (ii) make like Don Quixote and chase after impossible battles, you can take the third choice: If a change seems reasonable (or at least unobjectionable), find a way to go along with it. If the change doesn't seem reasonable, decide to either find a way around it or fight it. Few of these battles can be won, so you are better off in most cases to find a way around the change.

For example, if a word has lost its meaning, become a cliché, or for any other reason become unacceptable, I simply stop using the word. *Facility* has been used to mean everything from a shopping mall to an athletic field to a rest room, which means that it no longer has any meaning of its own. I stopped using *facility* in about 1966; even before I came to work for ASA–CSSA–SSSA.

Another word I don't use is *parameter*. While originally it meant an arbitrary constant whose value characterizes a member of a system, *parameter* now seems to represent any characteristic element of anything: "Political dissent is a parameter of modern life." (I never use the word *modern* in my writing either; but for other reasons, which are explained on page 34.)

A few years after I began refusing to use certain words (and thinking I was the only person doing it) the pastor of my church told me he did the same thing: "I never use the word *justice* in a sermon anymore. As soon as I utter that word, people form a picture in their minds of what justice means, then react emotionally to their picture instead of listening to my message. 'Justice!? Justice!? How can you talk about justice when . . . ?'"

Other language difficulties

I (and likely you, either) will never eliminate the differences in spelling and word usage between the versions of English used in the USA (and much of the rest of the world) and in Great Britain and its former empire. So I look for ways to make information understandable in either version of the language. Take the ambiguous word *billion*, for example. Where the numeral itself or scientific nomenclature is inappropriate, I suggest that people writing for an international audience use the words *thousand-million* or *million-million*, to express the number they mean.

A frequent correspondent once suggested that I use this column to teach people how to spell, and urged me to write columns making fun of the people who don't know how to spell. I refuse. First of all, I don't like to make fun of people for any reason. More importantly, a few hundred or even several thousand words from me will never reverse a dozen or more years of inadequate schooling.

Also, (and I realize many will disagree) I have never believed that spelling is the most important element in writing. A well-written piece with a few misspelled words (or national spelling variations) can be understood, while an incoherent collection of perfectly spelled words cannot. Of course the ideal would be to have both perfect spelling and perfect syntax and sentence construction in a paper. If I have to choose, though, I'll take construction and syntax ahead of spelling every time.

A competent editor can quickly and easily correct a lucid paper that happens to have a few spelling variations or misspelled words. That same editor would have a difficult-to-impossible time trying to make sense of an illogically jumbled and incoherent mass of words, even if every word was spelled correctly.

I *will* fight when there is some hope of winning. Instead of using this column to teach spelling, I use it and other efforts to try to teach authors to present their ideas understandably. If authors do that, any competent editor working with them can put the final touches on the presentation.

* * *

One place where there are rules (squishy rules, but rules, nonetheless) is the area of comparisons. In response to a question about it, I wrote the following column, which was first published in December 1994:

Comparisons of superlatives

Distinguished Professor W.A. Russell, at Iowa State University, writes: "There seems to be a problem creeping into our verbal and written language. This problem is when to use *more* vs. *-er* or *most* vs. *-est.* I think many persons are using *more* when *-er* is correct. Is there a simple rule? I have seen a statement that if the adjective is a single syllable then *-er* is correct, but I do not believe it is this simple."

Dr. Russell is correct; the rule that single syllables require *-er* and *-est* is basically accurate as far as it goes, but it is, indeed, too simple.

There *is* a simple rule that works for people who are familiar with using the language correctly: Use whichever sounds best in the particular situation, but that generally means *-er, -est* with shorter words and *more, most* with longer words. That rule alone may not be enough to help some of you, so I'll go into a little more detail, and then at the end give you a more all-encompassing rule.

The problem is that while there are some words that should be used with only one form, there are others that allow more flexibility. Consider *narrower* and *more narrow*: That board is narrower than this one. That doorway is more narrow than recommended for safe emergency egress.

Some absolutes

My favorite dictionary for such things, *Fowler's Dictionary of Modern English Usage*, contains these rules; not all of which I agree with:

All monosyllable adjectives take *-er* or *-est* (hard, sage, shy, etc.) So do all words in the following five categories:

a. Disyllables ending in certain letters:

-y	(holy, lazy, likely, etc.)
-le	(noble, subtle, etc.)

-er (tender, clever, etc.)
-ow (narrow, sallow, etc.).

b. Many disyllables with the accent on the last syllable: (polite, profound, etc., but not antique, bizarre, or the predicative adjectives afraid, alive, alone, or aware).
c. Trisyllabic forms of (a) and (b), above: (unholy, unlikely, ignoble, etc).
d. Some other disyllables in everyday use not classifiable under terminations, such as common, cruel, pleasant, and quiet (but not constant or sudden). Many others in this category could logically go either way: awkward, brazen, buxom, crooked.
e. Adverbs not formed with *-ly* from adjectives, but identical to those adjectives: (runs faster, sleeps sounder, hits hardest, hold it tighter). True also for some independent adverbs (*soon*, *often*).

My thoughts on the above

While Fowler is an excellent source, he is not the only one. Even in the list quoted above, I see several words that seem to me capable of going either way. What is wrong, for instance, with *more holy, more likely, more unlikely, more subtle,* or *more profound*?

The further one moves down through Fowler's list, the longer the words become, and the more they seem to be also acceptable with the use of *more* or *most.* This does not mean that I disagree with Fowler's rule, but rather that I think it is not a steadfast rule.

Some words sound strange with the *-er* but not with the *-est.* For instance, I prefer *more common* to *commoner*, particularly because that word has another meaning. I have no problem with *commonest.*, but *most common* does not bother me either. A similar statement could be made for *cruel; more cruel* sounds much better to my ear than *cruller*; but I have no problem with, and actually prefer *cruelest.* The same could be said for the word *pleasant*, but even here, I think the context can make a difference: (You are much more pleasant today than you were yesterday; did you get a good night's sleep? I cannot remember when I've had a pleasanter time.)

Another opinion

The book *American usage and style: The consensus* by Roy H. Copperud (1970, Van Nostrand Reinhold, New York City) also is more lenient than Fowler. It contains sentences such as: "Most monosyllables (like *loud, soon*) are compared by adding *-er, -est; loud, louder, loudest; soon, sooner, soonest. More loud* is not wrong; the form *louder* is preferable and predominant." As I wrote above: "Many disyllables are compared by adding *-er, -est,* but here there is more choice: *nobler, holiest, narrower; more noble, most holy, more narrow.* Longer words are seldom compared other than with *more, most. Beautifuler* is something a child still getting a grip on the language might say, but it would not ordinarily be written except by the unpracticed or the affected."

Copperud concludes his discussion with: "In general, the ear is the best guide to the formation of comparatives. When in doubt, use the form that sounds more agreeable, or agreeabler."

That sounds very much like the rule I gave in the third paragraph of this column. I promised you a more complete rule, and here it is; a rule of thumb that will be correct in most occasions: Short words generally sound better with *-er* and *-est*, mid-length words can generally go either way, and long words generally need *more* or *most.*

Shall I conclude with a silly example to stress the point of not using *-er* or *-est* with long words? "He is more antidisestablishmentarian than I" is preferable to, "He is antidisestablishmentarianer than I."

* * *

Supposedly the longest English word in the previous sentence serves as a lead-in to the following column, published in August 1995:

Big words: When to use, when not to use them

Can you agree with either of these statements?

- Scientists have no choice, they must use complicated and long scientific terms when they write.
- Scientific writing could be simplified.

Both are correct. Science is complicated and so are some of the words used in it. But the *necessary* large words aren't what make scientific writing difficult to read; scientist–readers know what those words mean. What causes problems are the *unnecessary* large words that authors often use between the necessary words. Scientists have more reason than other writers to avoid unnecessarily long and complicated words.

For example, the key phrase in the previous sentence is "unnecessarily long and complicated words." Two words in this phrase are fairly long (one has six, the other four syllables). I placed the phrase at the end of the sentence and lead up to it with relatively small words (four with two syllables; and one word, *scientists*, with three).

Had I built the sentence with unnecessarily long and complicated words, finding that key phrase would be like trying to find a lighted candle in a room full of lighted 100-watt light bulbs: "Owing to the complexity of their subject matter, scientists have ample reason to be cognizant of their readership and, to a greater order of magnitude than other authors, to therefore conscientiously construct sentences that eschew unnecessarily long and complicated words."

As short as possible

How large a word should an author use? The best answer to such a vague question was supposedly given by Abraham Lincoln when asked how long a man's legs should be: "Long enough to reach the ground."

If there is no better way to write about *sequential extraction analysis* than by using those words, by all means use them. If there is no better word to use in a particular situation than *approximately*, use it also. But in the latter case you can usually use *about*, which I recommend using whenever possible, especially when it appears in a sentence that also contains *sequential extraction analysis*.

Writing by formula

Perhaps you have heard about indices designed to measure the readability of a piece of writing. These generally count or ask an author to count the number of syllables in each word, the number of words in each sentence, and run these through a formula to calculate the paper's readability. Authors have told me they had poor results when they tried using these formulae on their own scientific writing. I'm not surprised.

The material in the old *Crops and Soils Magazine* was not difficult to read (it was deliberately too simple for most scientists, in fact), but it generally scored poorly on those readability tests. You see, the magazine needed to use multi-syllable words, such as *compaction, fertilizer, applicator, insecticide,* and *irrigation*, which presented no difficulty to the farmers and farm advisors who subscribed to the magazine.

The same could be said for scientific writing. There is no need to create a simple or short substitute for scientific terms. Instead, someone needs to recalibrate those readability tests for science and scientific writing.

Small words are powerful

Two steps to becoming a good writer involve (i) always using the most appropriate word, and (ii) realizing there is nothing shameful or childish about little words.

Sir Winston Churchill, Franklin Roosevelt, and John F. Kennedy all knew plenty of big words—alliances, miscalculations, encumbrances, negotiations, diplomacy, etc.—and used them when necessary or appropriate. They also knew a secret; unnecessary large words do nothing to enhance powerful ideas, and usually hide those ideas. When these men needed to communicate clearly and motivate people, they used small and powerful words.

During the Battle of Britain, Prime Minister Churchill told his nation (either directly or, according to an actor who many years later said he was the one who actually read the words because Churchill was to busy fighting the war to travel across town to speak them himself) via radio: "We will fight on the beaches, we will fight in the streets, we will fight in the alleys, we will fight in our homes, we will never give up!" Those words, and more like them ("the only thing I have to offer is blood, tears, toil, and sweat"), galvanized the nation and gave it courage to endure almost constant bombing, fend off a much larger and more powerful enemy, and win one of the most deadly battles of World War II.

Only three of Churchill's words about fighting (*beaches, alleys*, and *never*) had even two syllables. The rest had one syllable apiece. What do you suppose the outcome of the Battle of Britain would have been if Churchill had said, "In light of the perilous situation currently confronting our beloved nation, it is imperative that every citizen endeavor to the best of his or her respective abilities to establish priorities such that . . . ."? The war might have been lost before he finished reading that sentence.

How believable or motivating would Churchill's personal promise have been if he had said, "I assure all citizens and hereby utter as a solemn contract the promise that throughout this terrible ordeal I have already begun and shall henceforth continue to expend my entire allowance of energy, sympathy, musculature, and character"? (By the way, you may have thought Churchill's promise contained the more famous, "blood, sweat, and tears." Sometime after the original broadcast Churchill rewrote the phrase to make it even more powerful for history.)

During the Great Depression, incoming President Franklin Roosevelt rallied the USA by telling its people, "The only thing we have to fear is fear itself." A generation later, incoming President John Kennedy told the people: "Ask not what your country can do for you, ask what you can do for your country."

Two words in the sentence from Roosevelt's inaugural address, *only* and *itself*, contain two syllables apiece; the remaining eight words are one syllable each. The only two-syllable word in the sentence from John Kennedy's inaugural address is *country*, which he repeats once. The 15 other words are all single syllables.

There is nothing childish about properly used small words.

Be nice to your reader

I once worked on a daily newspaper in Dubuque, IA, a city on the Mississippi River. One spring, during a bad flood, a wirephoto came in that showed a head of state who was attending a meeting in Venice, Italy. It was a gorgeous picture of the man standing alongside the Grand Canal, with gondolas in the background. The editor did not publish the picture with story about the meeting: "We've had enough pictures of water in the paper today; the readers need a break!"

Give *your* readers a break. Use all the long words you need, but don't fill your paper with additional long words you *do not* need. Be like the politician who won a school board election with the promise he would "spend every necessary dollar, but not one penny more." Remember also that small words are not necessarily childish. Properly used small words will improve the communication of your facts and ideas.

* * *

From one extreme to the other, one of the major problems of scientific writing is the use of far too many abbreviations, as pointed out in this column from January 1993:

Too many abbreviations confuse instead of help

You are reading a manuscript and come across the abbreviation EPS. You turn to the abbreviation footnote at the beginning of the paper and see that the author uses EPS to mean embryo protein synthesis. This makes sense, you think, and read the rest of the paper with no problem.

Then you move onto the next paper in the same journal, again see

EPS, try to read it as embryo protein synthesis, and find it makes no sense. You look up the definition and find that the author of this paper uses EPS to mean equidistant plant spacing. It takes some effort to mentally shift gears, and reading this paper is not as easy as reading the first.

Abbreviations and acronyms are supposed to save space and make it easier for a person to read a paper. If they don't do both, they should not be used. (In case you have ever wondered, an acronym is an abbreviation that can be pronounced as a word. Thus, AID, CIMMYT, IRRI, and ICRISAT are acronyms and ASA, CSSA, SSSA, USDA, ARS, and NSF are abbreviations.)

A second problem with abbreviations or acronyms is when so many are used in a paper that it is difficult to keep track of them all. Eventually you get to the point where they all come together, like this sentence from an actual paper: "Decreases in HC and CL at 15 d PA and decreases for HC, CL, and LG at 32 d PA may be related to concurrent increases in NCS." Raise you hand if you think this is more readable than it would have been with words in place of the abbreviations.

Such examples occurred so frequently in Society journals that in 1990 the editorial boards of all journals agreed to discourage excessive use of author-created abbreviations.

The questions for authors, then, became (i) How many are too many? and (ii) How do I write a paper without using abbreviations?

Don't abbreviate at all

The simplest way to avoid using abbreviations is to delete them. They often are not needed. I read a paper recently that described the plot as "500 m above mean sea level (asl)." The term "above sea level" was not used again in the paper, so there was no need to have an abbreviation.

This is not to say that terms used only once should never be abbreviated. The only time I can think of when they might be suitable, though, is when something is better known by its abbreviation than by its whole term. Certain international centers, such as CIMMYT or ICARDA are better known by those acronyms than they are by their full names.

A few research techniques, such as ELISA and NMR, are at least as well known by their abbreviations or acronyms and should be used even if they appear only once within the text of a paper. Certain chemical compounds such as DNA and RNA fall into the same category.

Aside from those, even if a term is used five or six times in a paper, you still might be better off to simply write out the term instead of making up an abbreviation that is confusing to readers.

Avoiding abbreviations

If you are using the term embryo protein synthesis, see if you can simply use the word *synthesis* instead of making up an abbreviation. If the paper deals only with protein synthesis by the embryo, you probably could. If the papers deals with protein synthesis by many plant parts, you obviously cannot use only *synthesis*, but see if you could use something else, depending on the situation. For example, if you are writing about various compounds synthesized by embryos, maybe you could just use the name of those compounds.

For equidistant plant spacing, perhaps you could use "this spacing" or "this pattern."

Don't give a new meaning to a recognized abbreviation

The abbreviation C stands for the element carbon. Do not use it to mean *center* or *cotyledon* or any other term beginning with that letter.

Conclusion

Certainly there is a place for author-created abbreviations, but not nearly as prominent as it has been. Editorial boards and the professional editing staff at Headquarters have been instructed to eliminate as many as we can. You can help by not using them in the first place.

* * *

From too many large words and too many abbreviations, we move to simply too many words. This column was first published in August 1993:

Unnecessary words and (unnecessary) phrases

Experts say many of Edgar Allan Poe's short stories were so tightly written that a person could not eliminate a single word without harming the work. I don't think anyone could say that about most of my writing, and I know that it is not the case for much of the writing that crosses my desk each day. In this column, I will show how some typical sentences, most from scientific papers, can be shortened without changing the meaning. My first example is the title of this column; the word in parenthesis should be eliminated, so the title would read, "Unnecessary words and phrases."

Let's take a simple sentence that, to some, may seem tightly written. "I think that this effort will be very successful." Anyone familiar with the language will immediately see two often unneeded words—*that* and *very*. Without those two words, this sentence would read: "I think this effort will be successful."

The word *very* can be dropped from essentially every positive sentence with no loss in meaning. (I realize that the word *very* has a precise meaning in some definitions—very fine silt is different from fine silt, so I'm not writing about that. Just don't let it carry over into the rest of your writing.)

Very adds meaning to negative constructions: "It wasn't very successful." is not the same as "It wasn't successful."

One needs to use more care in dropping the word *that*. It is sometimes needed. The following sentence should have the word *that* between *me* and *how*. "A scientist at the National Soil Tilth Lab told me how farmers drive and inflate their tires is critical." If you try, though, you can eliminate the word *that* from 25 to 75 % of your sentences.

Similar to the word *very*, the word *located* can be eliminated virtually every time it is used. "The test plots are located near Urbana, IL." means exactly the same as "The test plots are near Urbana, IL."

The word *factor* is another often unnecessary word. Consider the sentence, "Environmental factors are critical in evaluating maize." It could also be written, "Environment is critical in evaluating maize." The word *factor* contributed nothing to the meaning of the sentence.

Another sentence, "Nitrogen is normally a key factor in achieving optimum cereal grain yields." Without the word *factor*, the sentence reads: "Nitrogen is normally a key in achieving optimum cereal grain yields." Something's wrong there, so change the word *in* to *to*. "Nitrogen is normally key to achieving optimum cereal grain yields." Again, the word *factor* contributed nothing to the meaning of the sentence.

Unnecessary phrases

I cannot count how many sentences have I seen in scientific papers that read something like: "It will be shown that crop yields can be improved." Eliminating the opening phrase leaves: "Crop yields can be improved."

I'm sure you have often seen a sentence like the following in an acknowledgment section: "We wish to thank Jonathan Glotzgahagen for his help in this study." Instead of wishing you could thank him, go ahead and do it (we don't mind). And because it comes at the end of a paper about a study, what else could it refer to? Taking out both unneeded phrases leaves: "We thank Jonathan Glotzgahagen for his help."

A phrase like "we will" or "we would like to" is about as unnecessary as "we wish to." For example: "In this paper we will demonstrate the importance of management." Removing that phrase leaves: "In this paper we demonstrate the importance of management." We haven't lost a thing so far, but the sentence has another unnecessary phrase. Getting rid of it leaves: "We demonstrate the importance of management."

Another unnecessary phrase is "the advent of," as in: "Before the advent of computers it was difficult to carry out statistical evaluations." The same sentence without it: "Before computers it was difficult to carry out statistical evaluations."

I cannot think of many scientists who *don't* do their work carefully, but I can think of many who want us to know for certain that they do: "After a careful analysis of the results, we determined . . . ." The sentence is just as forceful as: "After analysis of the results, we determined . . . ." Do you see another unnecessary phrase? (Hint; what else but the results would they analyze?) Taking it out, too, leaves: "After analysis, we determined . . . ." (If you wanted to get really picky, you could even argue that "after analysis" is unneeded. While this is a valid deletion, it seems to slightly alter the meaning.)

More words and phrases

The list of unneeded words and phrases mentioned so far in this column is by no means exhaustive. Among other usually unneeded words, one can think of such gems as *parameter*, *following*, and *facility*. Among worthless phrases are such gems as "it is shown that," "it is emphasized that," "it is a fact that," and "it is known that." Nobody would miss any of these if they were never used again.

Consider the many long phrases that could be shortened to a word or two. Instead of using "appears to be," use *seems*; replace "in the absence of" with *without*; and "higher in comparison with" with *more than.*

Although some scientists might believe, "Why say something in one word when you can use five?" I think most readers would prefer to have authors turn that phrase around.

If you want to learn a few more unnecessary words and phrases, take an objective look at your own writing. Each time I look at my writing I find many. (An earlier draft of that sentence read "Every time I look at my own writing I find a great many.") The first draft of this column was loaded with unnecessary words and phrases. I hope this final draft contains only the intentional ones.

* * *

It's not only big words that cause problems; from February 1999:

Little words, big problems

Sometimes the smallest of words can cause the biggest problems. Certainly they can elicit the most questions. I'm often asked, for example, when and whether to use the word *a*.

Rather than try to concoct an answer, let me show a few examples.

A, vs. The

If you write, "Dr. Martha Jones has served as director of the Soil Science Institute for 3 years," you are saying that she was the sole director. If, on the other hand, you write, "Dr. Martha Jones has served as a director of the Soil Science Institute for 3 years" you are saying she was one of several members of the board of directors of that organization.

At first glance it would appear that the next statement would follow the same rule. It does, but with a twist. "Dr. Andrew Anderson has been working as a research scientist in clay mineralogy for 10 years." You would use this construction even if he were the only scientist so employed on her particular campus or laboratory, because there are others in his discipline across the nation and around the world.

If you wanted to follow this rule to the letter, you would write, "Dr. Jane Peterson is professor of soil science at the American University" if she were the only such professor at that university, but "Dr. Jane Peterson is a professor of soil science at the American University" if she were one of several. More often than not, though, I see sentences of this sort written without the *a* even when there is obviously more than one person involved: "Dr. Jones, professor of soil science at Iowa State University, said . . . ."

How important is the difference between these little words? A disagreement between them in the authorizing legislation held up the construction of the John F. Kennedy Center for the Performing Arts in Washington, DC, for several months. The two alternatives:

1. "... it will be a memorial to President Kennedy in Washington."
2. "... it will be the memorial to President Kennedy in Washington."

Those who argued for the first version wanted to leave the option open for other memorials sometime in the future. Those who argued for the second, did not.

Nearly as profound a change can happen in scientific text. Consider the following sentence beginnings: "A reason for this is . . . ." vs. "The reason for this is . . . ."

Clearly, there is a huge difference in the meaning of the words mentioned thus far in this column, and in the meaning of many other words that only seem small. So before you ask the question, "Which should I use?" you first need decide, "What do I want to say?"

Now let's move on to some other small words that can cause problems for other reasons.

This

Often a person will write a sentence, then in the next sentence seem to refer back to something in it with the word *this*. The only problem is that unless the author is careful, the *this* could refer to several possible parts of the sentence, or even to something else not in the sentence. Let's say you write, "Dr. Smith is a soybean geneticist. He has been working in this area for 5 years."

I cannot tell from the context of the second sentence whether the author wants *this area* to mean geographical area, such as "Dr. Smith has been working at the National Laboratory in Los Alamos (or wherever the author happens to work) for 5 years," or "Dr. Smith has been working on soybean genetics for 5 years."

To be more clear, I would use one of the following to replace both confusing sentences: "Dr. Smith is a soybean geneticist who has worked at the National Laboratory in Los Alamos for the past 5 years." or "Dr. Smith has worked as a soybean geneticist for the past 5 years."

If you ever use *this* in a sentence, make it clear what you are referring to. The only exception to the previous sentence is in those rare occasions when you need to be deliberately vague, as I did a few paragraphs ago with "A reason for this is . . . ." vs. "The reason for this is . . . ."

We

As any of you who have read more than a handful of these columns will know, I strongly advocate the use of active voice and a great reduction in the amount of passive voice in scientific literature. Thankfully, I am glad to see such a reduction has been happening. One problem that can happen when authors take steps to eliminate passive voice, though, is that they sometimes use an excessive amount of the pronoun *we*. A paper that I recently edited had that pronoun in nearly every sentence. Here are four consecutive sentences from that paper:

"We obtained soil samples for this study from a site near . . . . This one of several sites we have previously studied. We selected this site because it is representative of the most prevalent ecosystem in this geographical area. We collected 19 soil cores . . . ."

While there is nothing wrong with this particular pronoun, overuse of it presents as great a problem as overuse of any other single word. One is tempted to come up with a terrible pun about the undesirability of we–weing on your readers.

How can those previous sentences be modified to eliminate some of the repetition? Actually many ways. Two of the most common are to make some of them passive and to combine some of the shorter sentences. Here is a combination of both performed on the four sentences quoted above:

"We obtained soil samples for this study from a site near . . . . This particular site, one of several we have previously studied, is representative of the most prevalent ecosystem in this geographical area. We collected 19 soil cores . . . ."

A vs. An

I'm also asked from time to time when to use "a" and when to use "an" in a sentence. I saved this for last because it is one of the easiest questions of this sort to answer. In this context the words have exactly the same meaning and the choice of which to use depends almost entirely on the first letter of the word that follows it. The idea is to eliminate a doubling-up of a particular sound [see also p. 157 of this book]:

a boiler vs. an autoclave
a car vs. an automobile
a crocodile vs. an alligator

One of the few of these pairs that is in dispute involves which of those two words should precede the words *historic* or *historical*. The problem is that these words can be pronounced with two slightly different levels of emphasis on its first letter.

I go along with the few remaining purists who write and say, "It was an historical event."

More and more persons, though, opt for, "It was a historical event."

* * *

Still more on problem words, from March 1999:

Useless, misused, or confusing words

You can usually eliminate the word *The* from the beginning of a sentence, but you may not always want to. You might as well eliminate it in any sentence like the following:

Questionable: "The increased use of [whatever] in [whatever or wherever] has produced . . . ."

Better: "Increased use of [whatever] in [whatever or wherever] has produced . . . ."

To stave off a host of rebuttals, my first sentence stated that you can *usually* eliminate, not *always* eliminate it. The word is perfectly legitimate at the beginning of sentences such as "The meeting was held at . . . ."

The word, *the* also is a convenient word to place at the beginning of a sentence that would otherwise start with an abbreviation:

Questionable: "EDTA testing revealed . . . ."

Better: "The EDTA test revealed . . . ."

(See more below on beginning a sentence with an abbreviation.)

Another often useless word is *did.* A creative writing instructor told me once that I should never use the word *did* in the way I'm about to show unless I italicized it:

Correct only if italicized: "Previous use of higher crop residue amounts *did* increase total soil N content."

Correct all the time: "Previous use of higher crop residue amounts increased total soil N content."

Beginning with an abbreviation

Should you ever begin a sentence with an abbreviation? Don't be too quick to say "no." I'll bet at least once in your life you have written a sentence that began with something like, "Dr. Jones says . . . ."

I've also seen newspaper stories that begin with something like, "NATO foreign ministers met today in Belfast."

Clearly, close to all knowledgeable persons occasionally begin a sentence with a few selected abbreviations and many use an occasional acronym at the beginning of a sentence. I can make a stronger case for using acronyms at the beginning of a sentence than I can most abbreviations. Acronyms are, after all, abbreviations that form a pronounceable word. They are often treated like a word in other ways, such as being pluralized or used with an apostrophe to form a possessive. So a sentence that begins, "CIMMYT's new wheat . . . ." is understandable and not objectionable except to those few who refuse to recognize the ongoing evolution of the English language.

I cannot remember using an acronym at the beginning of a sentence myself. If I have not, though, I likely will sometime soon, for I regularly allow it in papers I'm editing.

I tend to be extremely careful in my use of abbreviations at the beginning of a sentence, though. Aside from *Dr.*, I cannot think of any others I have used. For example, I would not write, "ASA announces the beginning of a new service to its members." Even with abbreviations that do not necessarily need to be defined, as the EDTA in the example above, I think it is better to insert a word like "The" before it or restructure the sentence to move the abbreviation inside.

Match nouns and verbs

In English, a singular noun always requires a singular verb and a plural noun always requires a plural verb. Don't be fooled by words that come between the noun and verb, they don't change a thing:

Wrong: "Total root elongation for 1994 and 1996 were . . . ."

Correct: "Total root elongation for 1994 and 1996 was . . . ."

If the sentence construction makes this unclear, I'll move the noun closer to the verb: "For 1994 and 1996, total root elongation was . . . ."

Word placement

Placement of words within a sentence can often make a sentence more (or less) clear, as was demonstrated in the previous example and will be again in this letter from a reader:

"In writing a literature review, decide what you *really* want to be the subject of a sentence, then place that subject near the beginning of the sentence. Too often, authors begin sentences with who did the work, instead of with what was found. Consider this example: 'Abercrombie and Fitch (1999) found that a six-pound trout, if not played properly, could break a ten-pound test line.' The sentence is fine if the author is writing a biography of Abercrombie and Fitch, but rarely is that the case. More likely, the author intended to focus on the trout, or the line, or perhaps the angler, or even the angling technique. I'm sure your readers can rewrite the sentence in several ways, with each version having a different emphasis.

"Several years ago, I made it against the law for my students to begin any written sentence with a reference, unless the reference is the clearly intended subject of the sentence. The quality of the students' writing improved immediately."

– Ron Taskey
Cal Poly Soil Science Department
San Luis Obispo, CA

[I hope you notice that I just demonstrated the way to mark a multi-paragraph quote. Place an opening quotation mark at the beginning of each paragraph, but a closing mark at the end of only the final paragraph.]

* * *

And still more on words; this one from June 1996:

Word use and misuse

This month I want to again consider words, and I'll begin with things that only sound like words. Maybe it is because I am in the business of putting words together that I have a keen ear for these things, but I hear them frequently. Their sound does to me something akin to what the sound of fingernails scraping across a blackboard does to other people.

Here's one at the beginning of this sentence: "Irregardless of what you might think, it isn't true." *Irregardless* isn't a word, the author should have used *regardless*.

"Evidentially there are some who do not believe what we say." The first collection of letters sounds to the untrained ear like a mighty impressive word, but it's not. The author should have used *evidently*. Attorneys use the word, *evidentiary*, which may be the source of the non-word and the reason why it sounds impressive to the untrained.

Then there is a class of words that might be called Archie Bunkerisms, named after a character on the old "All in the family" TV show. At least once in every show Archie Bunker would use a real word incorrectly. My favorite was when he said, "Let's sympathize our watches" when he meant, "Let's synchronize our watches."

A Madison politician during a campaign speech a few years ago said,

"We are wasting too much money on duplicity. We need to get rid of so many people doing the same job." *Duplicity* means a particular type of lying or cheating, *duplication* was the word he meant.

Of course, both Archie Bunker and the politician meant well. It is hard sometimes to figure out what the correct word might be in a situation. The pronouns *I* and *me* confuse many people, including even President Clinton. Which would be correct in this sentence? "He made that for John and (I or me)." To decide, say the sentence to yourself without the other person in it, then put the person's name back in without changing the pronoun:

> "He made that for me" becomes, "He made that for John and me."
>
> "I am going" becomes, "John and I are going."

Want another rule-of-thumb to help you make a word distinction? Here's one for when you need to use *either* or *neither* in a sentence and follow it with *or* or *nor*. If the first word begins with an *n*, the second word should also. Thus, "either–or" or "neither–nor."

English has many word pairs that sound the same but mean something totally different, like *pair* and *pear*, or *way* and *weigh*. Those pairs don't seem to confuse anyone, but another is probably the most misused pair of words in the English language; *its* and *it's*.

The word *it's* is a contraction for "it is." You would use it in a sentence like, "It's raining hard today." ("It is raining hard today.")

The word *its* is an entirely different word. It's a possessive.

It's easy to understand why the words are confusing. Normally to form a possessive, one adds an apostrophe and an "s" to the end of a word; "John's sailboat." There are two other possessives that do not use apostrophes, though, and David G. Rossiter of Fall Creek Software in Ithaca, NY, put them into the following rule that makes the distinction clear: "HIS, HERS, and ITS are all possessives, and NONE of them takes an apostrophe, because NONE of them is a contraction!"

You would use the possessive word in a sentence like, "I thank the board of directors for its strong leadership." And *I* thank David Rossiter for his contribution to this column. (Judging by the capitalizations and exclamation point, this misuse must bother Dr. Rossiter as much as nonwords, mentioned at the beginning of this column, bother me.)

A similar problem happens with the abbreviations i.e. and e.g. These aren't interchangeable terms, but are confused in a third to half of papers we receive that use these abbreviations. It is not always easy for a professional editor to figure out which the author meant.

The i.e. means, "that is." Think, for a moment, of a pathologist ruling that the cause of death was poisoning. The e.g. means, "for example." That would be like the same pathologist saying poisoning was one of many possible causes of death.

If you want to use the Latin abbreviations, use i.e. to define an exclusive term or list, and e.g. to give some examples. Better yet, don't use either. Many newer style manuals recommend use of the English equivalents instead. That way you can state explicitly what you mean and editors who work on the paper don't have to guess whether you meant what you wrote or perhaps didn't understand the terms.

If, after all of this, you still want to use the Latin abbreviations, here is a suggestion from Gundega Korsts, former Headquarters professional editor: Think of e.g. as "egg," and then say, "For egg-zample."

In case you are curious, Korsts supplies the origins of the two terms:

> The abbreviation i.e. stands for *id est*, meaning "it is" (literally) or "that is" (only slightly loosely), and thus almost exactly matches the English expression "that is." This is used to make an exact specification, a one-to-one match.
>
> The abbreviation e.g. stands for *exempli gratia*, which translates as "for a pleasing example." In English, we've dropped the "pleasing" part, taking it for granted that the example cited will be appropriate. The examples are a subset of possibilities, a some-to-more match.

I always hedge when someone asks me how concerned a scientist should be with style. If by style they mean whether to use S.I. or some other measurement unit, correct biological nomenclature, correct math nomenclature, abbreviating rather than spelling out chemical elements, etc., I tell them they should be very much concerned. If by style they mean spelling, grammar, punctuation, capitalization, etc., I tell them they normally need be only a little concerned. But in situations like the one just presented, when it comes to distinguishing between two similar terms that have substantially different meanings, then word-use style is essential to scientific accuracy.

Words are a complex and fascinating subject. While many of my columns have dealt with putting words together, this is the eighth devoted solely to words themselves. You must take words seriously. As Mark Twain said, "The difference between the correct word and the almost correct word is the difference between lightning and a lightning bug."

Taken too seriously, though, words can nearly drive you mad. When words are taken joyfully and with perhaps a grain of salt, they can be fun to use and even to play with. Just consider what poets, song writers, and comedians do with them all the time.

* * *

Ever find yourself in a trap of your own making? Here is some exposition on a few such traps, and how to extract yourself from them, that was first published in January 1999:

A few random thoughts and reader replies

What's wrong with being positive? I was editing a paper the other day, and came upon a sentence that contained the following phrase, "Most agronomists do not advocate the use of . . . ." I changed it to, "Few agronomists advocate use of . . . ."

• • •

Every now and then I find myself running into a self-made roadblock. I know what I want to write, but I'm not quite sure how to write it. I do not at the moment have an example from my own writing, but I believe the writer of the following sentence faced the same situation:

"The greater NaOH P_o following sesbania fallow than natural fallow on the Oxisol (Table 3), even though the total P recycled through biomass in the sesbania fallow was not greater (Table 5), may be the result of more rapid decomposition of litter from the sesbania than the natural fallow."

What's a person to do who has given birth to something like that? First step back, mentally at least, if not physically. Perhaps take a couple deep breaths, and ask, "What am I trying to say here, anyway?"

After doing that to the example sentence, here's the way I suggested that the author rewrite it:

"You will note in Table 3 that the level of NaOH P_o following sesbania fallow was greater than that from natural fallow on the Oxisol, even though the total P recycled through biomass in the sesbania fallow was not greater (Table 5). This increase may be due to more rapid decomposition of litter from the sesbania than from the natural fallow.

• • •

A reader writes: I did some studies on the effect of shade and plant age on nonstructural carbohydrate (NSC) composition of cotton. I measured the *concentration* or *content* of the NSC in different tissues.

The unit I used is gram of NSC per kilogram dry weight tissue (g kg^{-1} DW). During my writing, I used the word *concentration* in my paper. Some people told me that I should use *content*. When I use the word *content*, some reviewers suggest that I use *concentration*. This makes me confused. Which word (concentration or content) is better in my writing?

Example: "The NSC concentration (or content) of leaves was decreased by shade, but the concentration (or content) of squares (floral buds) did not change under shade."

Thank you very much for your help

— *Duli Zhao*
Department of Agronomy
University of Arkansas
Fayetteville

Dr. Zhao:

The words *content* and *concentration* are not interchangeable terms. Let's say you have two vessels; one of these vessels contains 1 mL of water and the other contains 10 mL of water. Let's say further that you dissolved one mL of sugar into each. Each solution would have a *content* of 1 mL of sugar, but the *concentration* of sugar would be roughly 10 times greater in the container with 1 mL of water. That is the grammatical definition of those two terms.

Soil scientists refrain from using *concentration*. They nearly always use *content*, because they say things really do not dissolve into the soil.

It is possible that the crop scientists who urged you to use *content* instead of *concentration* adhere to that theory as well. That some other crop scientists argued otherwise shows that the belief is not universal. I could not find any information about the terms in the *Glossary of Crop Science Terms*.

• • •

From another reader: "I am a biology student at the University of Alaska Fairbanks, and I was told that you are an authority on writing. My question is, when word processing a document, is it correct to put one or two spaces after each period between sentences?"

— *Jody Hoffmann*

Jody Hoffmann:

I'm not certain I can give an answer to that question that would be correct for every possible situation and every possible market. I can certainly tell you, though, that ASA, CSSA, and SSSA publications want only single spaces between sentences. I can also tell you this is the trend for every publication I know of, for two important reasons.

First, the extra spacing is no longer needed in most cases. In the days of manual or electric typewriters with the Courier type face, where a period took up as much space as the capital letter "M," it was difficult to tell where a sentence ended. People would insert an extra space at the end of sentences to help make that ending clear. Now that most people use word processors with proportional letter spacing, the end of a sentence is generally clear without any additional space.

Second, advances in technology mean that most typesetting is now done electronically, instead of manually by someone reading from hard copy. Electronic machinery gets upset if it gets two identical information pulses without intervening text.

This is true for tables as well. Even if you need to change your tab stops with every line, use only one tab between items in a table. (Of course, different rules apply if you use the word-processing table-making function. There you have no tab stops at all.) [Chapter 10 gives detailed information on the production of tables.]

* * *

And then there comes the question of jargon, which—in case you haven't heard—is perfectly fine *in its place*. From June 1992:

Of course use jargon, just use it correctly

During a speech on a college campus I asked, "How many had heard or read somewhere, 'don't use jargon'?" Every hand in the room went up.

Then I asked, "How many of you can give me a good definition for the word *jargon*?" No hands went up.

I suspect the response would be similar regardless of where I asked those two questions. People who try to promote better writing are being unfair when they do not define this term. If most people do not know what jargon is, how can they avoid it? More importantly, if they do not know what jargon is, how can they know for certain that they *should* avoid it?

Jargon has a two-part definition. It is (i) the specialized or technical language of a group or profession that (ii) sounds nonsensical and incoherent to everyone else. Every profession, trade, and group of people has its jargon. The word jargon, in fact, comes from a word that means "the twittering of birds." The implication being that birds can understand each other, but nobody else can understand them.

So, if you want to communicate with birds, twitter in their language. Thus, the title of this article—of course use jargon, just use it correctly.

Jargon is not imprecise language. For example, the phrase "moisture stress" is not jargon, it is imprecise language—it is used to mean a plant was stressed by either too much or too little water. If you ask five weed scientists what "preemergence application of an herbicide" means you will probably get five different answers because that phrase also is imprecise language and not jargon. Likely, depending on the profession involved, persons could give a great many more examples of imprecise language that are sometimes mistaken for jargon.

There are two different kinds of true jargon, (i) words that are specific to an entire profession or substantial group within that profession, and (ii) words that are specific to a much smaller group. As an example of the latter, if everyone in your laboratory calls a 250-mL Erlenmeyer flask a *kazonga*, then it is perfectly acceptable to ask a colleague in that lab, "Please hand me a kazonga." It would not be appropriate, though, to use that term in a lab on another part of campus or another state, or in a paper that would be read by scientists all around the world.

And, although the phrase "250-mL Erlenmeyer flask" would be perfectly appropriate in scientific writing, it might not be appropriate at a science day presentation at a local elementary school—"funny-looking" or "special kind of" bottle might be better understood by that audience.

The key to using jargon properly is to think about the audience; not just who they are, but why they are reading or listening to you. You might use one type of language to describe the toxicology of a pesticide to a physician who asks you for information on how to treat a patient who was exposed to it, and a far different sort of language to explain to that same physician what to do about mysterious spots on her houseplant.

Before you begin to attempt to communicate with someone, ask yourself if you are attempting to convey vital information or an interesting tidbit. Is it for formal publication in a scientific book or journal, less formal publication in a trade magazine or extension publication, or casual conversation in a hallway or shopping mall? Is it to a colleague who wants to repeat your work or to your child who asks, "What did you to today?"

Scientific journals are read by scientists and people studying to become scientists. All are intelligent, most know little or nothing about you or your work, and some have difficulty reading and understanding

English. If you think you don't need to be careful about how you present your material because, "Everybody will know what I mean," you will be incorrect. Most will not. People read scientific journals to advance their careers. Even if you forget everything else in this article, if you remember the contents of this final paragraph, your writing will improve.

* * *

I revisited this subject in February 1998. This is slightly redundant with the previous, but has enough that is new to warrant publishing both.

Specialized language fine, in its place

Every one of us—myself included—uses specialized language when dealing with peers, coworkers, or members of some particular groups to which we might belong. There is absolutely nothing wrong—and a great deal that is right—about such specialized language.

The problem comes when we need to communicate with someone who is not a part of that group.

Ask any plant pathologist (or human health-care worker, for that matter) what the word *vector* means and you will be told it is an insect or anything else that transmits a disease from one organism to another.

Ask any airplane pilot or air traffic controller what the word *vector* means and you will be told it means to give a direction, as in, "Please vector me to the nearest airport."

Ask any mathematician what the word *vector* means and you will get a third definition.

Ask most of the people in the world how large the number *billion* is and you will be told "a thousand million." Ask anyone with a British or Commonwealth background, though, how large the number *billion* is, and you will be told "a million million."

When I speak or write to an agronomist, crop scientist, or soil scientist about some aspect of writing or editing or typesetting or printing, I try to be careful to either not use the more esoteric terms commonly used in those circles or to define them in context. For instance, instead of telling someone who does not work every day with printers and typesetters, "Give me a sidebar on that," I would write or say, "How about explaining the details of that procedure in a brief article we could publish alongside the main text." Sometimes I add, "what printers call a sidebar."

Remedy when speaking

What is a person to do when he or she inadvertently uses a word the listener does not understand? The remedy while speaking is simple, first hide your frustration, then define your term.

At one of our annual meetings I was speaking about writing instructional material, and said something along the lines of, "Be factual but do not pontificate." Several people interrupted and asked me to define the final word in that sentence.

My first unspoken reaction, "Everybody knows what the word *pontificate* means." My second unspoken reaction, "Obviously these people don't know, so they must be stupid." My third unspoken reaction, "No, I know they are not stupid, but that word is obviously one they are not familiar with."

I'd be willing to bet I am no more arrogant than anyone else, and the thoughts and observations in the previous three paragraphs have gone through your minds from time to time when you use a—to you—common everyday word and your listener looks at you as though you have just said something in Klingon. (I'll define *Klingon* below in case you are not familiar with it.)

One problem with specialized words is that they so perfectly characterize what they mean that they are not easy to define in other words. After all, everybody (that we know, anyway) *knows* what the word means.

Take *pontificate.* It most likely comes from the same root word as *Pontiff,* which is one of the titles of the Pope. Under certain rare instances (twice, I believe, in the past two millennia) a Pope has made a solemn pronouncement defining a matter of faith or morals. These ex cathedra (from the throne) statements are considered by Catholics to be infallible (in a similar manner that a ruling by the Supreme Court is infallible—there is no higher authority on Earth to contradict it).

So one who *pontificates* speaks with great (verging on too great) authority. I remember saying once about someone who had just given me some bad advice, "He pontificates with too little knowledge." Thus, it is a more pejorative than admiring term.

To be consistent, I had better also define a word I used four paragraphs ago, *Klingon.* That is a language created for use by one group of characters in the *Star Trek* movies and is also used in the later *Star Trek* TV programs. That language has developed a cult following. A small group of persons (not including myself) speak the language to each other at conventions and other gatherings and some books (including the Bible) have even been written in or translated into the language. To the best of my knowledge this is the only instance of (limited) public acceptance of a totally made-up language. But since the total number of people who use the language is so tiny, I figured it would be a better language to use in that example than, say, Greek, which is both a cliché in sentences of that sort and a better known language, particularly among academics.

Remedy when writing

There is no instant feedback when you write, so you need to consider whether any of the terms you commonly use may not be universally known and understood. When you write, try to understand the background of the audience you are writing for. Then try to use terms you know they will understand. How can you do that?

The first step, before you begin to write, is to take time to familiarize yourself with the language used by your potential readers. If you are writing a textbook, for example, the ideal situation is to have taught several classes to that particular age- or ability-level group. If you have not had direct contact with the audience that will read the book, at least take time to study other textbooks written for the same group. Note the vocabulary used and try to use the same vocabulary in your book.

After you have done your research and written as best as you can at the level of your target audience, try if possible to show your draft to someone with a similar background as that intended audience. If you are writing a college freshman-level textbook, for instance, ask a student at that level (or an advanced high school senior) to critique it for you.

If you cannot do that, the next best bet would be to ask someone who regularly writes for that audience to critique your work.

In short, it all comes back to the same thing I have written many times in these columns; the only way to successfully communicate with an audience is to know that audience.

* * *

I was accused of being politically correct when I wrote this one, which was published in April 1993. No I wasn't, I was simply factual:

Writing your manuscript in gender-neutral terms

The leaders of our three Societies voted at the annual meetings last year in Minneapolis to encourage (but not require) use of gender-neutral terms in our publications.

"Gender neutral?" you might ask. "On top of everything else I have to make my scientific writing gender neutral? How am I going to do that?"

Actually, it won't take many changes. Most of the papers I see make no references to gender; there is no gender to test tubes, computer disks, field plots, or ANOVA tables. Still, there are times when gender will creep into a paper. For instance, a paper may contain a sentence such as, "When a farmer applies this herbicide he should be careful not to let the vapor drift onto nearby crops."

What's wrong with that sentence? Not all farmers are men. Similarly, not all scientists are men, nor are all secretaries and lab assistants women. Perhaps some of you have been in our Societies long enough to remember the portion of the annual meetings called the "Ladies' Program?" As more and more women became professional members of the societies, and other changes took place within society as a whole, this program evolved to "Spouses' Program" and then to "Companions Program."

There are a number of ways to make the sample sentence, above, gender neutral. The simplest and cleanest way to make the change is to simply pluralize the sentence: "When farmers apply this herbicide they should be careful not to let the vapor drift onto nearby crops."

One problem with pluralizing a sentence is to make certain all terms that refer to the subject agree. Otherwise, the sentence could have errors such as: "When anyone applies this herbicide, they should be careful not to let the vapor drift onto nearby crops." The subject, anyone, is singular; the verb, they, is plural (see more on this below). There are two correct ways to avoid this problem. One is to make the subject plural: "When people apply this herbicide they should . . . ." The second is to make the verb singular: "When anyone applies this herbicide, he or she should . . . ." (I try to avoid "he or she" because it sounds so awkward.)

There are people who think they can avoid using "he or she" in a paper simply by alternating the two pronouns. This technique works if you are careful, but can produce some strange images and often make it difficult to tell whether one or two people are involved: "First he did this, then she did this, then he did this, and finally she did that."

An even more awkward construction than "he or she" is "he/she:" When a farmer applies this herbicide he/she should be careful not to let the vapor drift onto nearby crops." Even though it is cumbersome, I have used "he/she" from time to time when I cannot think of a better way.

There is another correct way, and also an incorrect way, to deal with this problem. One is to stick with the "they." I do not recommend it, but increasingly see sentences like, "Everyone should exercise their right to vote." Secondly, more and more feminists are recognizing the centuries-old understanding that "he" and "his" in constructions of this sort is more of a generic term for a human being than referring specifically to one sex.

The most complicated solution would be to totally revamp the sentence: "Herbicide drift can be damaging to nearby crops." The main problem with doing this, as this example demonstrates, is that such radical changes often alter—at least to some extent—the meaning of the sentence.

Like the terms *spouse* and *companion*, there are similar gender-neutral words that are valuable to learn because they are so convenient to use. For example, the word *workmen*, depending on the precise meaning required, can be replaced with such words or phrases as *personnel*, *staff*, *work force*, or *workers*. Here is an example of this substitution:

Sexist: Workmen should not be allowed to reenter fields until 12 hours after spraying.

Neutral: Farm workers should not be allowed to reenter fields until 12 hours after spraying.

The somewhat archaic and slightly formal word "one" can be perfectly clear, direct, and occasionally useful in scientific writing: "When applying this herbicide one should be careful not to let the vapor drift onto nearby crops."

As well as the use of the word *he*, discussed above, is that other formerly acceptable words have also fallen out of favor, at least in the minds of many. *Mankind* and *man* used to be used to represent all people. Thus, in 1977, the Societies published a book titled *Crops and Man*. Fifteen years later, when the second edition of that book was issued, the decision was made not to update the title to reflect contemporary sensibilities. Care was taken in the advertising of that book, though, to use language that was acceptable to the time the second edition was released.

Instead of man or mankind, words such as people, humans, human beings, or the human species can often be substituted, depending on the precise meaning required.

Another way of achieving gender neutrality is simply to state the person's name. Dr. A.B. Jones can be referred thereafter as Dr. Jones with no need for anyone to know if Dr. Jones is a he or a she.

Some people may have to make a few initial mental shifts, but writing in gender-neutral terms is not necessarily awkward. Gender-neutral writing can be as accurate—sometimes even more accurate—than sexist writing. Writing that is accurate, unbiased, and not awkward is clearly a goal worth striving toward.

* * *

After all that seriousness about words it's time to have some fun. The rest of this chapter will have fun with words, while also conveying a surprisingly large amount of practical information you can use.

Some words about words

One of the first articles I wrote for the first newspaper I worked on after college began with the sentence, "[name] was hurt following an auto accident." I handed it to the crusty city editor and started to walk back to my desk on the other side of the newsroom. Before I got to it, the city editor shouted at me in a voice loud enough for all to hear, "Hey, Luellen, how far did he have to follow that accident before it finally hurt him?"

It didn't take too many embarrassments like that for me to begin paying careful attention to the meaning of words. To this day I will not use *over* when I mean *more than*, *lower* when I mean *less than* or *fewer*, and, of course, *following* when I mean *after* or *in*. (The *Dictionary of Modern English Usage* grudging allows *following* to mean *after, because of*, so I have thrown in the towel on that one and tolerate it in sentences such as "Following a good crop year we had a record harvest." I still do not like it, though, and rarely use it myself.)

One misuse that, for some reason, particularly bothers me is the word *feel* as a replacement for the word *think*. A person can feel happy or sad or hungry, but cannot "feel that breeders would be reluctant to ignore environment" or "feel that we ought to do it this way." It makes far greater sense to write or say that you "believe breeders would be reluctant to ignore environment" or that you "think we ought to do it this way." Check your writing and speaking: Each time the word *feel* is not immediately followed by a descriptive word such as *sad*, *happy*, *hungry*, *frustrated*, *tired*, or whatever, change the word *feel* to a word like *think*, *believe*, *contend*, or *propose*.

Similar, but separate terms

There are number of pairs of words or terms that look or sound alike, but have substantially different meanings. Take i.e. and e.g., for instance. The term i.e. means "that is to say," it introduces another way to say something: "My colleague is an idiot, i.e., dumb as a post." The term e.g. means "for example," or "for instance." "My colleague is an idiot, e.g., the other day . . . ."

I sometimes wonder if there are more than three or four people in the entire nation who know the difference between two substantially different words, flout and flaunt. Flout means to flagrantly go against—as to flout the law, or flout traditional morality. Flaunt means to make an ostentatious display of—as in, to flaunt one's new Ph.D. degree or automobile.

There are phases for the moon and phases in the growth cycle of plants and animals, but if something happens that does or does not faze you, and you need to write about it, spell it the way I just did.

If something doesn't jibe, it means it doesn't fit in well with what came before it. If someone doesn't jive, it means he or she is not in with the current popular culture.

Words upon words upon

Anytime you find yourself repeating a word—e.g., "we sampled a sample"—you can just about bet there is a better way to phrase that sentence. (This rule does not necessarily apply to titles of columns about writing.) The following example has three such repetitions; the word *root* appears three times and *diameter* and *sample* appear twice each in one difficult-to-follow sentence: "Additionally, the crop's root diameters were measured to determine maximum root diameter in each sample at the mid-December sampling to further examine the influence of seeding rate on root growth." [31 words]

Eliminating the redundancies and recasting the sentence makes it not only much tighter, but also far more understandable: "To further examine the influence of seeding rate on root growth, we measured the maximum root diameter of each sample taken in mid-December." [24 words] (If you prefer, you could reverse the two phrases and eliminate the need for a center comma: "We measured the maximum root diameter of each sample taken in mid-December to further examine the influence of seeding rate on root growth.")

I took this final example from an actual paper. All the writing in that paper was as wordy as the sentence just quoted and modified. If its author eliminated the repetition of words within sentences throughout that paper he could have reduced his overall page charge 25 to 30%.

For other examples of misused words, see those one or my associates or I highlight on your papers. Unlike the first editor I worked for, though, we'll always try to call them to your attention in a way that won't embarrass you.

* * *

I conclude this chapter on words with what has become the most enduring feature of both the columns and the newsletter that continues it, similar but separate words. Although I present it here as three columns, in actuality the middle one contains a great many more words than it did originally, because I moved into it words from several more columns and newsletter issues:

Similar but separate words

English has few perfectly synonymous words, which makes sense; there is little need for two different words that have the same meaning. Two that come immediately to mind are *car* and *automobile.*

Two cities in England are *Henley on Thames* and *Kingston upon Thames*. Both cities have two syllables in their names and are on the same river. I can see no difference between them that would explain why one city name uses the word *on* and the other the word *upon*. In other words, *on* and *upon* are synonymous.

Whether you use *on* or *upon* (or both) is a matter of individual taste. For what it's worth, although many people I know prefer *upon*, being a simple type, I use *on* about 95% of the time and only use the longer version for cadence or other aesthetic considerations. As to my choice between *car* or *automobile*, it depends on how stuffy I want to seem. I drive a car, but the Chairman of the Board is driven in an automobile.

To a child, the words *father* and *daddy* have exactly the same meaning, but the meanings of those words part slightly to significantly, depending on the individuals involved, as the child grows older.

Most word pairs that seem similar are actually quite different. *House* and *home*, for example, have vastly different meanings.

To orient you to where I'm going with this, I've just given two categories, the rare true synonyms and the many words that seem similar but are quite different. There is a third category, words that are similar, but slightly different from one another. Failure to make the proper distinction between such similar-but-separate word pairs immediately brands the writer as one who is not very familiar with the language.

Although a few of the distinctions between similar-but-separate words are cultural, most distinctions today seem arbitrary. The historical distinction between the two words usually came about via the particular language from which each of the words originated. If you would like to learn their history, look them up in the *Oxford English Dictionary*. In this column I will not consider the reason for the differences, only inform you which is which in today's American version of English.

Gray vs. Grey

The British spell the word *grey*, while Americans (with a few exceptions) spell it *gray*. The main American exceptions involve trade or brand names; for example, Earl Grey tea and the bus line *Greyhound*. As to the dog itself, I have seen it spelled both ways. Some writers, because the breed originally came from England, seem to believe it should still have the English spelling, and make it *greyhound*. Others, I believe the majority, spell it *grayhound*. A few make it two words, *gray hound*.

While there may be some ambiguity about the spelling of that breed of dog, there is none about any other uses of the word *gray*. If you ever need to use that word to describe a soil color or plant disease symptom or for any other reason, and you are writing for publication in the USA, please spell it *gray*.

Insure vs. Ensure

Use *insure* only when writing about a legal policy and *ensure* in all other instances.

"I have a policy to *insure* my life." "I eat good foods to *ensure* a healthy life."

"She bought a policy to *insure* her farm against flood damage." "She built a wall of sandbags to *ensure* her barn would not be flooded."

"Planting several different crops, instead of relying only on one, will help *ensure* economic survival."

"Taking 50, instead of five, samples will help *ensure* you have a representative sample."

Farther vs. Further

Farther is the word to use when you are expressing an actual physical distance: "The farm you seek is five miles *farther* down this road." "Seeds planted deeper into the soil than their counterparts have *farther* to grow before they reach sunlight."

Further is the word to use when dealing with philosophical or ideological distances: "That is worth *further* study." "He believes there is a solution to this problem. *Further*, he is ready to stake his considerable reputation on a research project to find it." "Furthermore . . . ."

May vs. Might

These two words can be used three ways, with and their meanings are different in each usage.

In the first of these three situations the words are synonyms and tense alone determines which you should use. Present tense: "She says she *may* complete her thesis today." Past tense (at the end of that day): "She thought she *might* complete her thesis today."

In the other two uses of this pair of words their meanings are different from each other, but with a twist: Although both words normally indicate a possibility, exactly how remote a possibility each word implies depends on whether one is using the language casually or formally.

In standard usage, *may* indicates a slight possibility and *might* indicates a more remote possibility, as in this government edict: "Any broadcasting station that airs more commercials than the code allows *may* be fined, and in extreme cases its license *might* be taken away."

Those two meanings switch in casual conversation. At a party when the host asks if he can get you anything, if you reply, "I *might* have another drink," you mean "please bring me another drink." But if you answer "I *may* have another drink later," you are indicating you do not want a drink now and have not made up your mind whether you will later.

Sewage vs. Sewerage

Sewage flows through a sewerage system.

That vs. Which

You will often hear these words used interchangeably in song lyrics, commercials, and news broadcasts, and read them in fiction. A few dictionaries and style manuals also have stopped treating these words differently. But those of us who deal with the publishing of scientific information often need additional precision and cling to the important distinction between them. An earlier edition (but, unfortunately, not the current edition) of the ASA–CSSA–SSSA *Publications Handbook and Style Manual* used an excellent example to demonstrate the difference:

- I do not drink water that is salty.
- I do not drink seawater, which is salty.

In the first sentence, the word *that* introduces a restrictive clause, a clause that must be included or the sentence would not be true. As evidence that the phrase is essential, note it is not separated from the rest of the sentence by a comma. In the second sentence the *which* introduces a nonrestrictive clause, separated from the rest by a comma. The second sentence would still have the same meaning if the comma were changed to a period and everything after it in the sentence was deleted. The nonrestrictive clause conveys additional, but nonessential information.

To help me use these words correctly I coined my *water closet* rule. The British term *water closet*, which I will not define here, is often abbreviated *WC*. The *W* (in word *which*) is always accompanied by a *C* (comma). When I mentioned my *WC rule* to a group of people a few years ago, several told me of their own mnemonic devices to help them use the correct word. Unfortunately, I did not write down or remember any of them, and cannot repeat them here.

Final comment

The world will not come to an end if you use *further* when you mean *farther*, or spell the word *grey* instead of *gray*, or misuse any of the other similar-but-separate word pairs described here. (With the possible exception of *that* and *which*. Misuse of those might bring about the end of civilization as we know it.) It is even safe to say that most people who read your paper (if editors do not catch and correct it) will not even realize you made a mistake.

But your misuse will become justification for someone else to think, "that person did it this way and it was published, so this must be the correct way."

And bit by bit, drop by drop, another distinction dribbles away to nothing and our language is depleted a little more. Take a hint from Smokey the Bear, "Only you can prevent language destruction."

Handling three sets of problem words

A reader wrote expressing concern that I had used the phrase "different than" in something I had written [not published in this book]. I believe it was in the phrase, "different-than-normal usage." He said he had been taught that *different* should always be followed by *from*, and wanted to know if I used the incorrect word or if he was mistaken about the rule. He was correct in that case; I used the wrong word.

But the word *from* is not the only word that can correctly follow *different*. There are, in fact, three words that can be correctly used to follow the word *different*, depending on the usage and context. These three words are *from*, *with*, and *than*.

Here is how Fowler's *Dictionary of Modern English Usage* makes the distinction (definitions are from Fowler; some examples are from me):

1. In the sense of to be different or to exhibit a difference, the word *different* should always be followed by the word *from*. "Sample A is different from Sample B." "Different-from-normal usage."

2. In the sense of having a difference of opinion, express dissent, or to dispute something, the word *different* is usually followed by *with*, but occasionally with the word *from*. "I have a difference of opinion with her about that."

3. "Different than" is sometimes preferred by good writers instead of the cumbersome phrase "different from that which." Two examples:
   a. He is using the word in quite a different sense than he did yesterday.
   b. The changes in the Commonwealth since the War ended have left Great Britain in a different position than she was in 1938.

In summary; when you compare objects, you should use "different from." If you are comparing ideas and opinions, you may sometimes be better off to use "different with." And in cases where you would rather not write out "different from that which," you may use "different than."

Fowler's dictionary, by the way, is a delightfully written compendium that is valuable to both beginning and experienced writers. My well-used copy was copyrighted in 1965 (Oxford University Press, New York and Oxford). I bought it new in 1965. It has recently been updated.

Who vs. Whom

Here's a simple rule to help you know whether to use *who* or *whom*: Turn the sentence around, and if the inserted pronoun becomes *he or she*, use *who*; if the pronoun becomes *him or her*, use *whom*.

Let's say you have just written the sentence: "I don't know whom will be going." Turn it around so that it reads something like "I don't know if he will be going." Since the pronoun is *he* and not *him*, you can immediately see that the you should have used *who* instead of *whom*.

The sentences that follow are all correct:

"The book is valuable to both beginning and experienced writers; the latter of who generally call it simply 'Fowler.'" Turned around, the sentence becomes, "He generally calls it 'Fowler.'"

"To whom shall I send this?" Turned around, it becomes, "Shall I send this to her?"

"Who is going to do this?" "Will she be the one doing this?"

"I am one of the last of the holdouts who sees a difference." "He sees a difference."

By the way, this rule will be less help to those who insist in writing in passive voice. The following sentences are not correct (the *who* is correct, because the pronoun should be *he* in each case). The sentences almost sound as though they are correct, and could lead the person who is not entirely familiar with the language to use the incorrect word: "The one who calls it 'Fowler' is him" (he calls it "Fowler"). "The one who sees a difference is him" (he sees the difference).

More similar but separates

After a recent column on similar but separate words readers sent me other word pairs that had given them problems. Here are some of them, with the distinctions to help you keep them straight. I'll also help you sort out four pair of words that only look similar. [Also, as mentioned above, I have inserted a great many more pairs, and a couple trios, of words into this column that were originally published in other columns.]

In vs. Into

Use *into* when you begin from the outside. From a deck, platform, or diving board, you would jump or dive *into* a pool. If you are standing in the water when you jump up and down, you are jumping *in* the pool.

If you wanted to be cute about it, you could write, "While standing *in* a pool I poured water from a bucket *into* that pool." (If it helps you to remember, if it is not already in the sentence, mentally insert the word *while* before the word *in*.)

More practically, you would pour a reagent *into* a test tube. The only way you could pour a reagent *in* a test tube would be if you either had an unusually large test tube or you were an unusually small person.

(I was once thoroughly disgusted by a sign, "Don't throw waste paper in the toilet.")

Tall vs. Height

Tall refers to distance above its surroundings of something connected to those surroundings. A building or tree is tall. A maize plant, then, would be 2 m tall, not 2 m high.

Height has two meanings. The more common meaning is distance of something above the surroundings it is not directly connected to. An airplane or a balloon is flying at a certain *height*.

So, using this definition, the only ways you could correctly describe a maize plant as being *2 m high*, would be if it were a most unusual plant or that it remained at that height for no more than a fraction of a second. In other words, similar to: "Debris from the explosion reached a *height* of 20 m." (But, "Debris from the explosion was later found on the roof of a nearby 20-m *tall* building.)

It is possible to use both *tall* and *height* in one description: "The base of the cloud was 3 km *high*, and the cloud itself was 1 km *tall*."

The other meaning of *height* refers to distance above sea level. A 2-m tall maize plant could be growing at a height of 3 or 4 km above sea level if it were in a plot a few kilometers west of Denver.

Unless specifically mentioned in the sentence, the term *height* is normally given in relation to the surroundings, not sea level. An airplane described as flying at a height of 1,500 m at the coordinates of Denver is assumed to be that distance above the city. A plane at a height of 1,500 m above sea level at the coordinates of Denver would be buried, not flying.

Since vs. Because vs. For

The word *since* means "from some time in the past to the present." Thus, it would be incorrect to write, "That has been going on since 1953 until this month." Three correct alternatives to that sentence would be:

- "That has been going on from 1953 until this month."

- "That has been going on between 1953 and this month."
- "That has been going on since 1953."

The word *because* means "the reason is that." It would therefore be redundant to write, "The reason the experiment did not work out is because the sample was tainted." To correct it, eliminate one of the words:

- "The experiment did not work out because the sample was tainted."
- "The reason the experiment did not work out is that the sample was tainted."

Although it is still acceptable, few writers use the word *for*. Those who avoid it say the word is somewhat archaic and not as strong as *because*. If you want to use *for*, the best distinction seems to be in cases where the cause-and-effect relationship is not as direct as it is with the use of *because*. Thus (from Bernstein's *The Careful Writer*):

- Correct: "It is a good play because it was written by Shakespeare."
- Correct: "It is a good play for it contains pathos, humor, and high drama."

In that sense, then, it would be incorrect to write, "It is a good play because it contains pathos, humor, and high drama." I cannot think of anyone who would enforce this rule.

Before I leave these words I must also make a confession, I hate to see *because* used as the first word in a sentence. Most sentences that begin with it are abysmally arrogant: "Because some idiot did thus and so, now all of us good people are forced to endure . . . ."

A person frequently needs a word that means both *after* and *because of*. A few people use *following* (and Fowler now allows it), but I do not like it, for reasons discussed in an earlier column. Instead, I use *since*: "Since Watson and Crick discovered the makeup of DNA, the science of genetics has taken a quantum leap."

That use, technically, is incorrect. But it is a practical construction used by many people that I believe one day will become acceptable.

If vs. Whether

While *whether* is normally used to introduce a noun clause, *if* often can be used correctly there: "They asked whether we would attend the banquet." "They asked if we would attend the banquet."

The exception to the above statement is when the noun clause begins a sentence. Then the word *if* can seem to suggest a question: "If we were coming to the banquet was the object of his inquiry."

Another problem with *if* is that it is commonly used in *if–then* constructions. When people see *if*, particularly at the beginning of a sentence, they expect to find a *then* somewhere within that sentence. This gets back to the subject of *disguised writing*, which I discussed in my October 1997 column (see Chapter 4). The natural tendency in the above sentence would be to expect something like, "If we were coming to the banquet, then we would have . . . [perhaps] packed nicer clothes, [or] brought more money."

The word *if* also should not be used if it injects an ambiguity, which can happen because the word also carries the connotation of *if and when*. So if you tell someone, "Call me if you get into trouble," the person might infer that you expect failure.

Imply vs. Infer

The former is a positive action of the speaker or writer, the latter is the judgement of the listener or reader. In the previous example, whether or not you meant to *imply* that you expected the person would fail, the person *inferred* it.

Duplicity vs. Duplication

These two words (and the three pairs that follow) technically do not belong in this column. There is nothing similar about them except their appearance.

The words *duplicity* and *duplication* are so radically different in meaning, in fact, that I could not believe anyone could ever confuse them. Then I heard a candidate for the local school board (of all offices) say something along the lines of, "There is too much duplicity in the school administration building. We have five or six people doing the same job."

What the candidate meant was "There is too much duplication . . . ." Duplicity means *lying or cheating*.

Vary vs. Very

Vary comes from the same root as *variable*. That alone ought to let you see the difference between these two (shall I say?) *very* different words. As I've mentioned in another place, try to avoid using the word *very*. It has its place, of course, in definitions such as *very fine silt*. The word adds essentially nothing to a positive sentence. In my above example, "two different words" and "two very different words" have the same meaning.

Aside from definitions, about the only place where the word *very* contributes anything to a sentence is in a negative construction: "It was not very successful" has a different meaning from "It was not successful."

Discreet vs. Discrete

The thesaurus on my word processor gives the following synonyms for *discrete*: *different, distinct, independent*, and *separate*. So you could use it in a sentence such as, "The plant is made up of four discrete components; roots, stalk, leaves, and reproductive parts."

Discreet, according to that same thesaurus, means *careful, cautious, circumspect, guarded, discriminating, judicious, politic, prudent, wise*, and *careful*. With all of that, a person might get the idea this is a quite admirable word. Not entirely true. In popular use today the word carries an underhanded connotation: "Sure the politician cheated on her husband, but at least she was *discreet* about it."

Desperate vs. Disparate

The word *desperate* means *critical, dire, extreme, serious, or urgent*: "A desperate situation." And, to quote Fowler's *Dictionary of Modern English Usage* "The word *desperation* once meant despair, abandonment, and loss of hope. Those have fallen by the wayside now, and the word means 'a reckless readiness to take the first course that presents itself when every course seems hopeless.'"

In other words, neither *desperate* nor *desperation* are likely to be used much in scientific writing.

Disparate, on the other hand, has more use for scientists. It means *different, dissimilar, diverse, unlike, discordant*, or *incompatible*.

Then vs. Than

I never considered the meaning of these two words to be close and was shocked the first time I came in contact with a person who confused them. The person, a cartoonist who had submitted work to the old *Crops and Soils Magazine*, became rude and belligerent when I called his word-use error to his attention.

I know now from my own work and from talking to teachers that many others, particularly younger people, confuse these two words. Perhaps the confusion has something to do with the "see-say" instead of the phonics way of learning words; these words *do* look almost identical.

At any rate, *then* pertains to time. First you get your degree, *then* you teach. After I get my next paycheck, *then* I will pay this bill. To help keep these two words straight you might associate *then* with the number ten, a number on a clock, which is therefore related to time.

Than is a comparison. This is better *than* this.

There vs. Their vs. They're

You could correctly use all three of these words in a sentence like: "They're going there to receive their award."

Think of *there* as a locator. If you drop its first letter you have another locator, *here*. Use *there* in sentences like, "*There* goes a great person," or "Park your car *there*."

Their is a possessive. "That is *their* car," *Their* research data were destroyed by fire." To help you remember *their*, drop its first letter and you have *heir*, someone who will one day possess all that you own.

They're is a contraction for *they are*. The sentence in standard English, "They are coming today," would be stated in conversational English as, "They're coming today."

Affect vs. Effect

These are probably the two most-often confused words in the English language, even though every authoritative book on the subject will tell you there is clear distinction between them. Let me quote two example sentences from Fowler's *A Dictionary of Modern English Usage* (1965 Oxford University Press, New York and Oxford):

This will not *affect* (change) his purpose.

This will not *effect* (secure) his purpose.

An example by Theodore M. Bernstein in *The Careful Writer* (1965 Atheneum, New York City), is similar: "The two-party coalition *effected* the passage of the tax program, which will *affect* almost every taxpayer."

Part of the reason these words can be confused is that they can be used both as nouns and verbs. As a noun, the word *affect* has a narrow psychological meaning, and is pronounced with the accent on the first syllable. Aside from that, the noun to use would virtually always be *effect*.

Bernstein, in the previously cited book, writes: "Despite their quite different meanings, these two words tend to be confused. As verbs, *affect* means to influence or have an effect or bearing on, and *effect* means to bring about, to accomplish, to execute."

Fowler, in the previously cited dictionary, agrees: "These words are not synonyms requiring differentiation, but words of totally different meaning, neither of which can ever be substituted for the other. *Affect* (apart from other senses in which it is not liable to confusion with *effect*) means to have an influence on, to produce an effect on, concern, effect a change in: *effect* means bring about, cause, produce a result in, have as a result."

Here are some more examples from Fowler:

These measures chiefly *affect* the great landowners.

It does not *affect* me.

It may seriously *affect* (that is, injure) his health

A single glass of brandy may *affect* (alter for better or worse) the prospects of his recovery.

A single glass of brandy may *effect* (bring about) his recovery.

Bernstein, cited above, suggests the following mnemonic device: "When you mean execute—a word beginning with an 'e'—the word you want is *effect*, another word beginning with an 'e.'"

Like vs. As

I begin this section with some history from Bernstein, cited above: The word *like* is most often found in older English texts with the word *as*: "Like as a father pitieth his children, so the Lord pitieth them that fear him." (Psalms 53:13). Here *like* is an adverb with the meaning *in the same manner. As* performs the function of a conjunction. But early in the history of the language, *like as* became felt as a unit, and *as* began to disappear since people began to believe that *like* expressed both the meaning and the function all by itself.

More recently the word *like*, for some reason, has fallen into disfavor among many grammarians who say it is perfectly acceptable for a person to write "Runs while the gazelle runs," and "Runs after the gazelle runs," but not "Runs like the gazelle runs." The word in the latter example, the grammarians say, should be "Runs as the gazelle runs."

Commenting on this example, Bernstein writes: "There is no logical reason why *like* should not be regarded as a conjunction; on the contrary, there are sound logical reasons why it should. And yet those objections by grammarians, reasonable or unreasonable, are a force in the language as it exists, as contrasted perhaps with how it should exist. This does not mean that they are eternally entrenched." He cites examples, particularly from advertising, to show that the usage is seeping into popular culture, "Winston tastes good, like a cigarette should."

But whether you or Bernstein or I agree with them, here are the grammar rules (quoted and paraphrased from Bernstein's book) concerning the use of *like* in sentences:

1. *Like* is permitted when it precedes a noun that is not followed by a verb. "He plays chess like an expert. She acts like a lady. In such constructions, *like* is equivalent to the prepositional phrase, *similarly to.* A gerund, which is a verbal noun, may follow *like* in, "Walking stimulates circulation like swimming." But the word following *like* must be a true gerund, not merely have the appearance of one, as in this incorrect sentence: "The new Russian state did not look like withering away as predicted."

2. *Like* and *as* are not interchangeable. *As* cannot be used instead of *like* in places where the verb is suppressed or its implied presence is not strong. Thus, the following sentence is not proper, "The people behaved as children." Not only that, Bernstein adds, "it sound as hell."

3. Aside from some idiomatic expressions such as "The car looks like new," and "They cheered like crazy," *like* cannot stand for *as if.* Therefore, the following sentence is not proper, "He spent money like it was going out of style."

4. One of the most common misuses of *like* is in comparisons of things that cannot be compared. "Like ancient Athens, democracy flourishes in the United States." Ancient Athens is not *like* democracy. Correction: "As in ancient Athens, democracy flourishes in the United States."

Whereas vs. While

According to Fowler, cited above, *whereas* is properly used only in contrasts.

Bernstein, cited above, lists one proper, one acceptable, and one improper use for the word *while*:

Properly, *while* is used to mean "during the time that," ("He had the radio playing while he read.")

Acceptable, although with less universal sanction, *while* is used in the sense of *although* or *but.* ("While a brilliant student, he generally has no feeling for mathematics.") When the word *while* is used in this sense, the *while* clause precedes the main clause.

Improperly used, in the judgement of most stylists, *while* sometimes is made to serve as a weak substitute for *and* or a semicolon: "Dr. Klepper is president of CSSA while Dr. Cassel is president of SSSA."

Compliment vs. Complement

Your words may *compliment* another scientist, but your findings may *complement* that person's work or discoveries.

To give a *compliment* is to give congratulations. Other synonyms include acclaim, adulate, applaud, commend, enhance, and extol.

When you *complement* something, though, you complete it or add to it. (You can use the word *complete* as your mental clue to remind you which word is which.) Synonyms for *complement* include: adjunct, contingent, supplement, addition, extra. Here are both of those words used correctly in one sentence: "She complimented me for complementing her work."

Appraise vs. Apprize

When you *appraise* a person, situation, or thing you analyze, evaluate, or judge it. When you *apprize* a person, you give notice to, tell, or inform that person of something. Example: "After the athlete appraised the umpire's work unfairly, the umpire apprized the athlete that he was ejected from the game."

Imminent vs. Eminent vs. Immanent

Imminent is the one word of these three most likely to be used in scientific literature. Synonyms include approaching, close, coming, ensuing, forthcoming, immediate, and inevitable.

Use eminent if you want to extol someone (the word is usually used to describe a person). Synonyms include acclaimed, august, celebrated, distinguished, famous, foremost, honorable, illustrious, and influential.

Immanent is one of those words that are fun to know but that's about it. I have never seen it used. Instead, you are far more likely to see its more common synonyms, including inborn, ingrained, inherent, intrinsic, and sometimes internal and natural.

Orient vs. Orientate

Both words mean exactly the same thing; to get into proper position with respect to one's surroundings or circumstances. ("He spent his first day on the job trying to orient himself to his duties." "The three-day program is designed to orientate students to the goals and challenges of life at a major university and the location of the buildings on campus.")

So what is the difference? One sounds more impressive and officialese than the other. So if you want to sound pedagogical, important, and stuffy, use *orientate.* But if you prefer to use simple and precise expression, use *orient.*

Continuing vs. Continual vs. Continuous

The stress of the first of these three is on the absence of interruption or cessation. The second and third seem at first to have the same meaning, but they do not. The light emitted by the sun is continuous, but it reaches any single point (between the Arctic and Antarctic Circles, anyway) on a continual basis—that is, every day, but not 24 hours per day.

Theodore Bernstein, in *Dos, Don'ts & Maybes of English Usage* (1997, Random House, New York), uses all three in the following sentence: "The tension on the Soviet-Chinese border has been *continuing* for several years with *continual* minor outbreaks and *continuous* patrolling by both sides."

Adverse vs. Averse

Although both mean roughly the same thing, the first is far stronger than the second. If you are averse to using something it means you are disinclined or reluctant to use it. If you are adverse to using it, you loath it, you are hostile to it, you just cannot stand it.

Cited vs. Sited vs. Sighted

The first comes from the stem citation, and is properly used in the following sentence, "I feel proud that he cited my paper in his landmark paper."

The second comes from site, or position. It's most common use is in a sentence such as, "She sited her country retreat on the top of the highest hill in the area."

The third refers to vision. Thus, "From her new country retreat she sighted many rare and endangered wildlife."

Lend vs. Loan

Although most people in the USA seem to think these are synonymous, careful writers and speakers here (and most people in the rest of the English-speaking world) know they are not.

Lend is the word to use in all but business dealings. After all, the famous speech from *Julius Caesar* did not start out, "Friends, Romans, countrymen, loan me your ears."

So if you see an advertisement that reads, "If we can't fix your car right away we'll loan you one," chuckle at its poor use of the language, but don't copy it.

And while *loan* is a business term, it is not the preferred verb in this sentence, "The bank loaned the business $250,000."

Oral vs. Verbal vs. Written

The first and last of these three are unambiguous. The first refers to spoken words. "We had an oral agreement" means we agreed to something after we talked about it, but we have nothing in writing.

Written agreement means precisely what it says, the agreed-to words were written down and can, thus, be verified.

The meaning of verbal agreement is much fuzzier. Obviously the it means the people involved agreed to something with their words, but there is no way to tell if those were written or spoken words, or even something more esoteric, such as sign language.

Compared with vs. Compared to

The first one is the one to use in virtually every situation. The only time you use *compared to* is when you are comparing substantially different things: "Compared to a dinosaur, my computer is not at all old."

But when you are comparing one crop with another, one season with another, one season with a 50-yr average, one pesticide with another, one tractor or other implement with another, one computer program with another . . . well, I believe I've made my case. If not, I'll restate the first sentence. In virtually every situation you are likely to encounter in just about everything you write, *compared with* is the phrase you would use.

The same is true with the use of the word compare in a slightly different phrase, as in: "When you compare these results with those of Jones and Smith (1976) . . . ."

Same as vs. Identical

When you're interviewing a person for a position, 5 years of experience might count the same as an advanced degree, but the two are not identical. Frozen orange juice may be the same as fresh squeezed, but the two are not identical.

I would equate *just like* with *same as*, which means I disagree with Bernstein, who in *Dos, Don'ts, and Maybes of English Usage* (1977, Random House), equates *just like* with *identical.*

Sit vs. Set

You can set the stage, set your hair, and set a pencil on a chair. But you cannot set on a chair and a house cannot set on a hill. In each of those cases the word should be sit. There are a couple exceptions, the sun sets and concrete sets. In other words, except for the named exceptions, the word *sit* is a transitive verb, which means that it conveys an action from one word to another in a sentence—from the subject to the predicate, for those who wish I would use some grammar jargon from time to time.

Hence vs. Therefore vs. So

The thesaurus on my computer lists *hence*, *therefore*, and *thus* as having the same meaning. Then, its "related words" section for those three words lists *accordingly*, *consequently*, *ergo*, *so*, and *wherefore.*

All the above are used to introduce a logical conclusion. Which you use depends on a number of things, including where you are publishing your material and your own background.

Stuffy professorial types like to use *ergo*, whereas attorneys are more likely to use *hence*. I personally lean more toward *therefore* and *thus* in my ordinary writing, *accordingly* or *consequently* when I want to impress someone, and *so* in my conversation with friends.

So if you ever see me begin a sentence with one of these words, you can gain a sense by my choice of words the level of formality or informality I am trying to convey in that sentence.

Raze vs. Raise vs. Rays

I'm not certain why someone wrote and asked me about these three words either. There is nothing at all similar about them except their sound. In fact, the first two are direct opposites. Raise is sometimes used in the sense of construction—raise a barn. Raze nearly always is used in reference to demolition of a building.

I do not believe I need to explain the meaning of the word *ray* to readers of this book.

Fane vs. Fain vs. Feign

I'm not sure why a reader asked me about these, either; there is nothing similar about these words except their sound; they rhyme with each other.

Fane is an old word for church. I once knew a priest named Kevin Fane, which always seemed to me to be an appropriate name (in the same vein as Dr. Glass being an optometrist.)

Fain is an archaic word for happy or pleased or willing or obliged.

Feign is the only one of these three words still in use. It is a synonym for fake or phony or pretend: "At the conclusion of the boss's joke, the employees feigned amusement."

Vane vs. Vain, vs. Vein

You might think this trio of words does not fit into this category either. I'm guessing the reader who asked about them wanted to know which of these is the proper word to use in a sentence such as "In a similar vein, we should consider . . . ."

Yes, vein is the word to use in that situation. You can come up with a number of reasons to explain that terminology, but the one I like comes from the analogy of a vein of minerals in the earth: coal can be found in this vein and in similar veins.

Vain comes from the word *vanity*, which originally meant futility or waste of time, and is the meaning to be given in the Bible passage "Vanity or vanities, all is vanity." Although the root word now means *conceit*, the derivative words still retain the original meaning.

Vane, as in weather vane, is something that changes with the wind or, in the case of a few mechanical devices, can be used to change the direction of a fluid or gas flowing through a system.

Unaware vs. Unawares

Is it ever proper to use the second of these two words? Surprisingly, yes, occasionally. The word means *unexpectedly* or *suddenly* or *by surprise*. For example: "That caught me unawares."

The rule of thumb to remember is that this word can appear only at the end of a sentence. If the word is to appear within a sentence it should always be *unaware*.

A proper phrase would be "I was unaware that she had been granted tenure," not "I was unawares that she had been granted tenure."

Bond vs Bound

No doubt some of you are saying, "Why, there is no similarity between these two words." And for most of their definitions you would be correct. But both have a connection with binding agreements, though, making them perfect candidates for inclusion in this section.

Bound comes from the same root as boundary and bind. Thus, one meaning of the word implies a restraint that can be imposed either by oneself or from the outside (duty bound).

Bond, which comes from the same root as binding, as in a band or cord used to tie up something, can be used in the context of a binding agreement.

Generally *bond* is used more in the strictly legal sense, as in *performance bond* or *bail bond*, and *bound* is more of a private morality, as in "I am bound by our agreement."

Use vs. Utilize vs. Usage

Those who want to sound impressive shun the use of *use* when at best there is no need to and at worst they shouldn't.

The proper use for *utilize* is when something can be put to profitable use: "They utilized the manure as fertilizer." An improper use of the word (from an example by Theodore Bernstein): "He utilized a screwdriver to clean his fingernails."

Use *usage* only when describing a traditional practice. In other words, the following example, also from Bernstein, is incorrect: "Authorities have no reliable data on the effects of marijuana usage." The correct word would be *use*.

This is not just a matter of using the shorter word whenever possible, it is a matter of using the *correct* word all the time.

Literally vs. Figuratively

This is another pair of words that are not at all similar, but are misused by so many people that they need to be explained here.

Literally means it actually happened in the way described. *Figuratively* means the person is using the words as a figure of speech. In fact, that might be a good way to remember the difference between these words.

Let's say someone said, "His debate opponent literally buried him with arguments." The only way that could be done would be if the opponent put those arguments on paper and then buried the person with that paper.

Little vs. A little

When the letter "a" is used at the beginning of a word it normally reverses the meaning of a word—*moral* vs. *amoral.* Placing the word *a* before the word *little* acts in somewhat the same way, but not exactly, and depends to a great extent on the meaning given to the word *little* in the particular sentence. The specific example is when the word *little* is used to mean *without* or *none.*

When you say someone gives you *little help*, you have just said in a polite way that the person gave you essentially no help at all. If, on the other hand, you said the person gave you *a little help*, you are acknowledging that the person was helpful to you.

Likely vs. Liable

Long-term readers will remember my commentary on ensure vs. insure, where the first is used in everything but the strictly legal sense and the second is used solely in legal policies.

This pair of words is something like that, but with an extension. As well as in the legal sense of liability, the word liable can also mean something akin to the forecasting of something unpleasant.

Only the most died-in-the wool anti-progress person could correctly say, "We are liable to get our new computers tomorrow." Those who range from indifferent to looking forward to the event would say, "We are likely to get our new computers tomorrow."

It would be correct, though, along with strictly legal uses of the word, to say something like, "The bosses new policies are liable to cause resentment."

Incredible vs. Incredulous

The former describes a thing or an action, the latter describes reactions to it. "When asked their reaction to the incredible feat two members of the audience said, "I felt incredulous."

Some letters from readers

An earlier column of yours dealt with similar-but-separate words. Another pair often misused, *momentarily* and *in a moment.*

Have you ever listened to a flight attendant's announcement of the pending arrival? "We will be arriving in [city] momentarily."

The comedian, Shelley Berman was a master at making fun of grammatical errors in common usage. In response to the flight attendant's announcement above, Berman remarked, "Well, I hope we'll be there long enough for me to get off!"

Another word pair often misused by people whose first language is other than English is *upon* and *about.* Example: "I must think upon the concept before I make my decision."

Number three, *to* and *and.* "I will try and harvest the plots this week."

—*Robert M. Millaway*
Tampa, FL

My reply:

How delightful to be reminded of Shelly Berman. I haven't heard his name mentioned in years, but some of the party records and books he produced still cause me to chuckle when I remember them.

As to your other two examples, *on* and *upon* can be used interchangeably, but as you implied, neither can be used in place of *about.*

And I agree with you that *to* and *and* certainly cannot be used interchangeably either.

Sir:

Enjoy your "fine tuning" column in *Agronomy News.* As an immigrant in Canada, I have a hard time with the way Canadians use *bring* versus *take.* But maybe I am the one that's wrong here. For example: If I am on my way to the office, I would take my briefcase with me, rather than bring my briefcase with me. If I am in the office, I would remind myself to bring my briefcase from home tomorrow. I would not remind myself to take my briefcase tomorrow.

Please clarify this sometime—for the benefit of our whole country.

—*Tom Francis*
Novartis

My reply:

For the benefit of the whole country? Wow! That's an awesome burden. [Before I deal with his question, please note that Dr. Francis used the word immigrant correctly in his letter. In case anyone needs a way to remember when to use that or emigrant, just think of the words import and export. Dr. Francis is an immigrant to Canada and an emigrant from what ever nation he moved there from.]

First, the easy definition; *bring* can be used in a way that does not imply movement, but rather to produce: "The changes in Federal research grants will bring benefits to a number of scientists."

The word usage gets much more complicated when dealing with movement. The distinction depends on the location of the person writing the words. The clearest way to depict this is when the writer and the carrier are two different persons: "When you come to my office, bring your manuscript with you." (When the person heads in the direction of the writer, he or she *brings*.) "When you leave, take my advice with you." (When the person heads away from the writer, he or she *takes*.)

The same applies, but is not quite as obvious, when the person doing the carrying is also the person doing the writing. "When I leave home in the morning I take my briefcase with me." "When I come to my office in the morning I bring my briefcase with me."

In other words, Tom Francis is correct in his use of these two words.

Bernstein, in *The Careful Writer*, presents two additional uses for *bring* and one additional one for *take* as well. The next three paragraphs are a direct quote from his book:

Bring to a head: This phrase is rather repulsive: To bring to a head means to suppurate, or cause pus to form. Here are a few substitutes: to "bring to a climax," to "bring to a boil" (this is a different kind of a boil), to "crystallize," to "reach a crisis," and to "precipitate."

Bring to an end: Wasteful. End by itself suffices. Variations on this, also wasteful, are "come to an end" and "put an end to."

Take place vs. occur: Although dictionaries define take place as meaning to come to pass or occur, usage seems to favor confining it to that which is prearranged or scheduled. Occur (or happen) is used for that which is spontaneous or accidental. Therefore, a sentence such as the following would be incorrect: "The crash took place at about 8 a.m. as the three-car train neared Catanzaro." A crash does not *take place* and a commencement does not *occur*.

Sir:

Which is correct? "Our experimental design was a randomized complete block consisting of . . . ." "Our experiment design was a randomized complete block consisting of . . . ." Please explain.

[name withheld by request]

My reply:

Both are correct, but have different meanings. Which you use will depend on what it is you want to say. Is the design of your experiment experimental? If so, then you would write, "The design of our experiment was itself experimental. Instead of the way Scientist et al. (1993) established the design protocol, we modified it in the following ways . . . ." If that is what you did, then your first sentence is correct; you used an experimental design.

If you used a standard design for your experiment, then your second sentence would be correct. I'm guessing the second version is what you want to write, and suggest three ways to make it a little less ambiguous:

- "The design of our experiment was a randomized complete block consisting of . . . ."
- More simply, "We used a randomized complete block design, consisting of . . . ."
- Or, in passive voice, "The design used for this experiment was a randomized complete block, consisting of . . . ."

Chapter 8

Titles

The first, or among the first words of your paper that people read are the words of its title. The title, in fact, is one of the things that helps people decide whether they want to read your entire paper. As such, titles are extremely important. Here's what I had to say about them in November 1996:

Titles: The good, the bad, and the incoherent

If you have you ever tried to find someone named *Jones* or *Smith* in a large-city telephone directory you can appreciate what readers go through when they scan journal tables of contents looking for a paper with a title that begins, "Effects of," "Theory of," "Influence of," "Maintenance of," or *anything* of.

As well as being difficult to find in a table of contents, such titles are trite and weak. One could almost state as a rule of thumb that any title that has *of* as its second word could, and probably should, be improved.

In this column I'll show you how to revise those and other problem titles. The examples used will be titles I've collected from ASA–CSSA–SSSA papers during the past three or four years. I'll suggest at least one alternative to each. You may disagree, but I believe my suggestions are superior to the originals.

The final two sections of this column and chapter will give my thoughts on how to write a good title.

Place important words first

Placing the most important word at the beginning of your title will do far more than simply help people find your paper in a table of contents, it will induce more people to read your paper.

Few journal readers have the time to browse, so they look for specific subject matter. When readers have a limited amount of time and every title begins with "Effect of . . . " readers may not recognize every paper they should read. Those same readers would have no problem finding the subject matter of papers whose titles begin, "Soybean physiology . . . ," "Soil drainage . . . ," "Corn breeding . . . ," or other such key words.

Here's how to change titles so you place the key words first:

Original: "Effects of magnesium on cation selectivity and structural stability of sodic soils."

Depending on which of these words is the most important, a better title might be:

Suggestions: "Magnesium affects cation selectivity and structural stability of sodic soils."
"Sodic soils structural stability and cation selectivity affected by magnesium."
"Structural stability and cation selectivity of sodic soils affected by magnesium."
"Cation selectivity and structural stability of sodic soils affected by magnesium."

Original: "Theory of minirhizotron root directionality observation and application to wheat and corn"

Suggestions: "Minirhizotron root directionality: Theory vs. observation and their application to wheat and corn"
"Corn and wheat root directionality theory vs. observations in a minirhizotron"
"Root directionality theory developed from minirhizotron observations"

Original: "Effects of flooding on pH of rice-producing, acid sulfate soils in Indonesia"

Suggestion: "Flooding alters pH of rice-producing, acid sulfate soils in Indonesia"

Original: "Maintenance of yields and soil fertility in nonmechanicized cropping systems, Bolivia"
Suggestion: "Yield and soil fertility maintenance in nonmechanicized cropping systems in Bolivia"
Original: "Infiltration and redistribution of organic liquids in layered porous media"
Suggestion: "Organic liquid infiltration and redistribution in layered porous media"
Original: "The upper limit of solute dispersion and soil hydraulic properties"
Suggestion: "Solute dispersion and soil hydraulic properties: Their upper limit"

Remove weak words

The following four titles begin with more important words, but three retain the word *of* early in the title and all have other weak or unneeded words. I eliminated unneeded words and replaced low-impact words with more forceful ones:

Original: "Preferential transport of nitrate through soil columns containing root channels"
Suggestion: "Preferential nitrate transport through root-channeled soil columns"
Original: "Axisymmetric transport of water and solute underneath a disk permeater: Experiments and numerical model"
Suggestion: "Axisymmetric water and solute transport under a disk permeater: Experiments and numerical model"
Original: "Pore-scale network modeling of compaction and filtration during surface sealing"
Suggestion: "Modeling compaction and filtration networks at the pore scale during surface sealing"

Insert strong words

Sometimes instead of eliminating unnecessary words, one must insert necessary words:

Original: "Unsaturated hydraulic conductivities of fiberglass wicks and designing capillary wick pore-water samplers"
Suggestion: "Unsaturated hydraulic conductivities of fiberglass wicks used to design capillary wick pore-water samplers"

Original: "Unsaturated hydraulic conductivity from transient multistep outflow and soil water pressure data"

Suggestions: "Unsaturated hydraulic conductivity calculated from transient multistep outflow and soil water pressure data"

"Calculating unsaturated hydraulic conductivity from transient multistep outflow and soil water pressure data"

Original: "Inverse method to determine soil hydraulic functions from multistep outflow experiments"

Suggestion: "Inverse method uses multistep outflow experiments to determine soil hydraulic functions"

Make pairs look like pairs

The next two titles are from a pair of papers, but that fact is not clear from their titles. The addition of a single word makes the linkage more clear:

Original 1: "Water-stable aggregates and organic matter fractions in conventional- and no-tillage soils"

Original 2: "Aggregate-protected and unprotected organic matter pools in conventional- and no-tillage soils"

Suggestion 1a: "Water-stable aggregates and organic matter fractions compared in conventional- and no-till soils"

Suggestion 2a: "Aggregate- protected and unprotected organic matter pools compared in conventional- and no-till soils"

If the above change was not significant enough for you, turn the titles around:

Suggestion 1b: "Conventional- and no-tillage comparisons: Water-stable aggregates and organic matter fractions"

Suggestion 2b: "Conventional- and no-tillage comparisons: Aggregate-protected and unprotected organic matter pools"

Now it's your turn

How do you write a good title? I cannot answer that question quickly, because before you can think about *writing* a title you need to understand the purpose of a title. The title of a scientific paper in some ways is similar to a billboard along a highway. Each has only a second or two to say enough about the product (paper) to get a person interested enough to want more. Also, each has about the same upper word limit (about 12 words), although billboards often give their message in as few as two to four words. Titles for scientific papers that are too short usually do not work. In fact, I have yet to see a title for a scientific paper with fewer than 10 words that was satisfactory, and would say six is the absolute lower limit.

Analyze the words on a billboard or two that attract your attention and see if you can figure out why they attracted you. Likely it is because the words were either powerful in themselves or they were put together in a unique or powerful way.

Here's an innocent-looking line I read on a billboard nearly 40 years ago and never forgot: "There is no match for electric heat." The first word of this line is vague, its verb is the weakest of all, and the line has only one powerful word, *match.* Yet that sentence is incredibly powerful because its author properly arranged that one word among six other carefully selected words. This sentence says electric heat not only is warmer and more economical than all other energy sources, with electric heat you also won't have pilot lights that need relighting, nor will you risk fires, explosions, or terrible dangers and disruptions to your family. Many readers would add, "Right, and there's no smoke or smell either, which means a nicer home for my family."

Even if you disagree with it, you have to admire the above sentence. I do not believe it is possible to improve it. Making it longer would dilute its impact. Shortening it to "Electric heat has no match" puts a stronger word at the beginning and eliminates the weak verb, but also eliminates the marvelous double meaning of the central word and makes the statement harsh and blatant. Changing the verb to, "Electric heat needs no match," knocks out the other part of the double meaning and destroys the original's finesse.

Another billboard I remember was one I saw while driving an unairconditioned car across a desert. I felt hot, dusty, windblown, and thirsty as I looked to the side of the road and saw, "Wouldn't a cold [beer] taste good right now?"

I am not recommending that scientific authors write titles that look and sound like billboards, because while there are similarities there are also many differences. One major difference is that you normally see only one billboard at a time, while there are many titles listed together in a table of contents.

To see what I'm recommending for scientific titles, study my suggested changes to the titles in the first part of this paper. Like a billboard writer I knew I had only a few words to work with and tried to get the most out of those words and their arrangement.

Give them lots of thought

Now for the question, "How does a person write a good title?" The answer is, "With great care and thought." Most titles are (and should be) compiled from some or all of the key words in the paper. Do not just shovel those words into the available space, though, carefully consider the arrangement of those words, plus the verbs and other words you add to them, so you get the most possible impact.

Key words can occasionally be esoteric and cumbersome. If you have tried and failed to come up with a reasonable title using them, consider using other words instead. In Chapter 7 of this book is a column on the evolution of the English language and the dilemma authors face as they try to decide whether to go along with or resist that evolution. After wrestling with those and other key words for some time, I threw them all away and instead borrowed the words from a Kenny Rogers song, "Knowing when to hold and when to fold."

Magazine editor J.K. Drake had written an article on recycling cardboard boxes by grinding them up and using them for part of the roughage portion of a cattle ration. After exhausting all possibilities for titles using those words, she threw them all away and wrote, "Box lunch for beef."

While neither of the above titles would be appropriate for a scientific paper, the idea is sound. I advise beginning editors to take about the same amount of time writing a title as they did writing the rest of the paper. Authors of scientific papers probably could not afford to spend that much time on their titles. But I urge you to give your titles more time and thought than you have until now. I think you will appreciate the results and I suspect your readers will also.

Chapter 9

Abstracts

Perhaps the most important part of any paper is the abstract. That statement is true because many scientists read the abstract to help decide whether to read the entire paper. Thus, if your abstract does not capture the readers' attention and compel them to read further, most of those scientists will not read your paper. To quote from a statement you'll see later in this chapter, "Lavish your attention" on your abstracts. They may be the only hint most readers get of the sort of work you are doing.

I begin the second of the two columns reproduced in this chapter with, "An abstract is an interesting thing. It is attached to the beginning of a scientific paper to help people decide whether to read the paper. An abstract is also published without the rest of the paper in places such as secondary and indexing journals."

Those two statements demonstrate the dual nature of an abstract, and in part explain why it is difficult to write one well.

Unfortunately, there is another duality as well; two entirely different kinds of literature both go by the name *abstract*.

The first is the kind just mentioned. A totally different sort of an abstract is prepared ahead of a paper presentation at a scientific meeting. It is unfortunate that these two totally different forms of literature have the same name, because it has caused confusion in the minds of at least a few scientists, as this column from May 1997 explains:

Two kinds of abstracts

Anytime a member tells me that something in a Headquarters publication is not consistent within itself I take notice and try to correct the error as soon as possible. Thus, my ears perked up a couple years ago when a member said to me, "You need to re-do that sample abstract you publish. It doesn't follow its own rules."

"Really?" I asked. "Tell me more. What rule does it break?"

The member replied, "The rules say not to use statements like 'will be discussed,' but the sample abstract contains several such statements."

I could not remember any such statements in the sample abstract, but was in no position to argue with the person, because it had been a few months since I had read it. I promised to look into the matter.

I pulled out my copy of our *Publications Handbook and Style Manual*, read the sample abstract, and I saw no statements like the person complained about. "Maybe he was thinking of a publication from some other Society," I said to myself and set the matter aside. A couple years later someone else brought me the same complaint. How could two people have read something in the sample abstract that was not there?

Annual meeting abstract

My confusion was resolved last year when I looked at the sample abstract that is sent to persons who will present a paper at our annual meetings. The annual meeting is handled by a separate department at Headquarters. I am not involved with the preparation of that department's materials. Also, I rarely give papers at our annual meetings, so do not have a reason to read those materials.

The sample abstract in that packet contains the phrase "will be discussed" twice and the phrase "will be covered" once. Is this inconsistent with our style manual? Not at all. An abstract for an annual meeting paper is an entirely different form of literature from an abstract for a scientific journal paper or book chapter. Their purposes are so substantially different from each other that it is unfortunate they are both called *abstracts*. [Note, I'm observing, not a suggested name change.]

Annual meeting abstracts are akin to a *preview of a coming attraction*. This type of abstract is usually written several months before completion of the paper, and often before completion of the research. In fact, these abstracts sometimes are written before the research is begun.

A scientist writing a preview of a research project cannot know its results, so has no choice but to write that the findings will be presented when the paper is given. Similarly, any consequences, theories, and discussions drawn from that research cannot be known (at least in full) at the time the abstract is written, so the author must again write that those will be discussed when the completed paper is presented.

Scientific paper abstract

Scientific paper abstracts are summaries of published papers. This type of abstract is normally written after completion of the paper and is designed for two specific purposes. First, the abstract is published along with the paper for busy people to read first to help them decide whether to read the full paper. Second, the abstract is also published in several ancillary publications, so persons who will never read the journal can learn what is going on in the rest of scientific literature. A side function of this second role is to attract persons to read our journal or at least to contact the author to get a reprint of the full paper.

The scientific paper sample abstract (the one that appears in our *Publications Handbook and Style Manual* and a few other places) was selected more than 10 years ago by a committee that studied all of our publications for about a year looking for the best possible sample abstract.

The names of the authors were masked throughout the selection process so the judging was done solely on the quality of the abstract itself. As it turned out, the winner was written by a young scientist near the beginning of her career. When we saw how well this scientist, J.L. Steiner, could write, we added her to the editorial board of one of our journals. She has since served on several editorial boards and has risen to prominence in the Society and the profession. Her abstract that became our sample abstract is a model of simplicity, clarity, and completeness, and an excellent model for anyone who is writing a scientific paper.

Do not confuse the two

This is the time of year our members get a chance to study the sample abstract they received along with other materials to help them prepare a paper for our annual meetings. As you study it, please do not think we erred in our preparation of that sample abstract. We did not.

I would as unhesitatingly recommend its use as a guide for preparing an abstract for an annual meeting paper as I would recommend our other sample abstract as a guide for preparing a scientific journal paper.

Don't you get the two forms of literature confused, though. It would be nearly impossible to write a scientific journal type abstract for an upcoming annual meeting paper and inappropriate to write an annual meeting type abstract for a scientific journal paper. They are entirely different types of literature.

* * *

As a result of the above column, when the ASA–CSSA–SSSA *Publications Handbook and Style Manual* was revised in 1998, a more thorough discussion of the subject was presented in it. So I hope this matter has been resolved for good. Another question that frequently comes up about abstracts is the one presented in the title of the next column, which was published in July 1993:

What goes into an Abstract vs. what goes into the text

An abstract is an interesting thing. It is attached to the beginning of a scientific paper to help people decide whether to read the paper. An abstract is also published without the rest of the paper in places such as secondary and indexing journals. Thus, the abstract will be seen and read by 50 to 100 times more people than will read the scientific paper.

While all scientists know these facts, they don't all take them into account when they write their papers and abstracts.

In essence, an abstract is a miniaturized version of the paper; complete within itself. It should include an introductory statement, the rationale or hypothesis statement for the research project, an outline of the objectives or the specific reason for the research project, a methods section that tells how the research project was conducted, a results section, and a conclusion statement.

An abstract should never contain a phrase directing people to the paper. Statements such as "will be discussed" or "will be explained" are worthless to people who read the abstract in a secondary publication.

The framers of our style manual spent many hours discussing the amount of repetition to allow between abstract and text. They developed these basic premises: (i) The abstract will be read by more people than will read the full paper, and (ii) nobody will read the full paper without first reading the abstract.

From these premises, they developed the overall guideline that everything important in the paper must be repeated in the abstract, but important things in the abstract do not necessarily have to be repeated in the paper.

Among the specific rules that flowed from that rule:

- Plants and other biological species and soil descriptions must be identified by their scientific names the first time they are mentioned in the abstract *or* in the text. When such names are given in the abstract, they do not need to be repeated in the text.
- Similarly, chemical compounds, commercial trade names, etc. must be identified the first time they are mentioned in abstract or text, but not both.
- An unnumbered footnote is required at the beginning of the text to define all author-made-up abbreviations used in the paper. Any of those abbreviations that are used in the abstract must be defined parenthetically there, too.
- An abstract should never contain footnotes or table, figure, or literature citations because they will not be available to persons who read the abstract in a secondary journal.
- "Equations, formulae, and obscure abbreviations and acronyms are inappropriate in an abstract." (Quoted from page 12 of the *ASA-CSSA-SSSA Publications Handbook and Style Manual*).

Putting this information to use

What does all this mean in the practical world? For one thing, lower page charges, if the journal you are writing for assesses them. This is because if you define a term in the abstract, you do not need to repeat the definition in the text. If a conclusion is clearly stated in the abstract it normally is not necessary for that paper to have a separate "Conclusions" section in the text; a couple sentences at the end of the "Results" or "Results and Discussion" section will normally suffice.

Both of the above rules are occasionally violated in our journals. One recent paper had definitions for a term in both the first sentence of its abstract and the first sentence of its text. (We tried to delete the definition in the text but the author would not allow it.) Other papers sometimes have a conclusion in the abstract and a conclusions section (If a scientific editor leaves it in, we do not second-guess him or her.)

Other abstract details

The Societies impose a strict length requirement on abstracts—150 words for Notes and 250 words for full papers. The reason for these limits is that abstracts longer than this limit may be truncated, often in mid-sentence, by the publishers of secondary journals.

Of course, words vary substantially in length. A more accurate measurement is the number of characters. The managing editor of a secondary journal referred to above told me they assume an average of six characters per word. If your word-processing system can give you a character count, limit the abstract for a full-length paper to 1,500 characters, with a maximum of about 1,750 characters. The comparable numbers for a Note is 900 characters, with 1,050 as your maximum.

Abstracts have additional requirements. They should provide an informative account of the research project in impersonal, noncritical language. Named soil types should be fully described; plants and other living organisms mentioned should be described by their binomial scientific names. (Any of these mentioned in the abstract need not be repeated in the text.)

Abstracts should also be specific. Instead of saying that the results for the new technique were better than that of the old, say how much better with a measurable number or percentage.

[If you are writing for a journal published by someone other than ASA–CSSA–SSSA, check out their requirements and, in the event of differences, heed them instead of those written here. Essentially all publishers adhere to the 250-word limit.]

A few final words

Your abstract is your only chance to tell readers about your work. Lavish your attention on it. In my newspaper and magazine days I would advise a trainee to spend as much time writing a headline or title as they did writing the article itself. The reason was completely practical; if the title did not catch the readers' attention, they would never read the article. This might be a good thing to keep in mind when you write an abstract.

To repeat the essence of this column: An abstract must be complete within itself; important things in the paper must be repeated in the abstract, but important things in the abstract need not be repeated in the paper.

Chapter 10

Tables and Figures

After the title and the abstract, the next thing most people look at in a scientific paper are its tables and figures. These are important enough that I devoted four of my original columns to the subject—including the second one of the original series and one of the first of the newsletter that superceded the original columns. Here is the first one, from April 1992:

Unnecessary 3-D figures incorrect and confusing

Authors are submitting more and more three-dimensional figures to our journals and books. While 3-D figures are the only way to present some data, many other data are hidden, confused, or presented incorrectly in a 3-D figure.

Figure 1 is an example of a false 3-D figure. There are three problems with a figure of this sort:

1. The figure is scientifically incorrect; it contains a Z axis with no data on it. The only reason this figure is presented in 3-D is because the computer software had a shadowing feature that made it possible. This shadowing feature should have been turned off as the figure was prepared.
2. Figures like this make it difficult to align the top of any one bar with the reference point so as to tell the height of that bar.

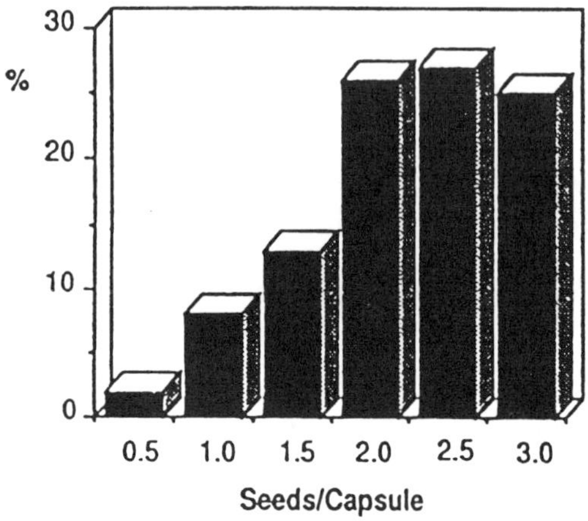

Fig. 1. Example of a false 3-D figure.

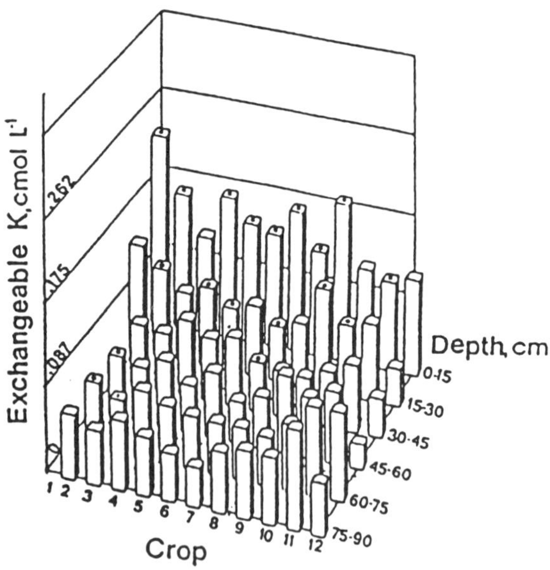

Fig. 2. Example of a confusing and obscured 3-D figure.

3. Figures like this can be visually confusing because some of the points and planes seem to be above and below others; much like the optical illusion art of M.C. Escher.

Figure 2 is an example of another problem that can happen with a 3-D figure. It is a true 3-D figure, with data on the Z axis, but those date are confused and obscured.

1. The figure has so many bars of substantially different height and obscured bases that it is impossible to tell which bar belongs in which row.
2. Another problem with this sort of figure is that shorter bars in back rows are hidden by taller bars in front of them. The figure in this example has so many rows with (seemingly) varying numbers of bars in each row that it is impossible to tell whether some of its bars are hiding other bars.

The information in Fig. 2 could have been presented much more clearly in two dimensions.

If the paper you are writing or reviewing contains one or more 3-D figures, take another look at them and make certain they are presenting the information in the best possible way.

* * *

Three columns later, July 1992, I came back to the subject:

What sort of info belongs in captions?

We received a letter not long ago from a friend and former Society officer about an inconsistency in the insert, "Become a more successful author," that we publish from time to time in all the Societies' journals.

The statement in question reads, "It is necessary to repeat some of the information from the text in the captions for tables and figures because readers generally study the tables and figures before they read the text. In the text, refer to tables and figures, but don't repeat them." The latter sentence is similar to one in the "Instructions to Authors" for *Agronomy Journal*: "Do not repeat information in the text [that has already been] presented in charts or graphs."

What captions must contain

The application of a little common sense will quickly cause this apparent inconsistency to vanish. Let's take a fairly simple table to illustrate my points. [To be honest, I cannot remember whether I made up the information in the following table or borrowed the table from a paper I was working on at the time. The original source does not really matter; the information in it should be considered generic. The important thing is that it is used here to explain how to describe information such as that presented not only in this, but in any table.]

What sort of caption would you write for a table like this, and how would you refer to it within the text of your paper?

	Rainfall	
Month	1991	Normal
	----- mm -----	
June	312	940
July	683	1394
August	1283	1321
September	1397	978
October	1318	688
Total	4993	5321

The first and most obvious question a reader will ask is, "What does this table show?" This table clearly shows the amount of rainfall during 5 months somewhere in the world, compared with the normal rainfall, presumably for the same time and place. The reader cannot tell from this table where these measurements were taken or why (although the paper's title and abstract may have presented that information).

A caption along the lines of, "Rainfall distribution during the growing season," gives no additional information, except, perhaps that the work was done in the Northern Hemisphere.

The typical caption for a table such as this in most agricultural journals would read something like: "Rainfall during the second year of a soybean experiment at Clayton, NC." I certainly would not fault such a caption (although in the section immediately following I'll show you one I like even better). First I want to deal with another matter.

Too many authors have the notion (in part because too many editors require it) that the purpose of a caption is to simply repeat every heading from the table. Thus, we find monstrosities such as: "Yield of grain, protein, oil, carbohydrate, and K of 'Hobbit', 'Elf', and 'Gnome' soybean grown in 1993, 1994, and 1995 in Centralia and Urbana, IL."

A similar monstrosity from a soils paper might read: "Measurement of S, Mg, Mn, P, and K in the sand, silt, and loam fractions of a sample of Muscatine silt loam soil collected in southeastern Iowa."

Neither of the captions in the previous paragraphs do a thing to improve reader understanding of the tables that would appear beneath them because they give no additional information about their tables. Readers get exactly the same information simply by looking at the table. Even if those captions were deleted, readers would still have exactly the same amount of information.

What text must contain

I'll come back to captions in the next section, but first I want you to consider what should be said in the text about the sample table on page 240 (as a representative of all tables).

First, it would be stupid to simply repeat all the table's information in the text, because the reader has already read and studied the table before even looking at the text. Still, you need to say something about it in the text, otherwise there is no point in having the table with the paper.

What I like to see is something like, "Rainfall for the season was near normal, but distribution was not (Table 1)," or "Rainfall distribution was unusual, dry early in the season and wet in the fall (Table 1).

What captions might contain

I have increasingly seen a new type of caption in tables. Instead of merely restating the contents of the table, or giving enough of a description so someone who has not read the text can have a rough approximation of what is contained in the table, these captions present the significance of the data. For a number of years I have been lobbying (with various degrees of success) for a caption for the sample table above along the lines of, "The weather at Clayton, NC, during the second year of our study was too dry for good second-crop soybean establishment and growth, and too wet for good drydown and harvest."

Some persons have advocated going even further, "The weather at Clayton, NC, during the second year of our study was too dry for good second-crop soybean establishment and growth, and too wet for good drydown and harvest, yet our test yields were near what would be expected with optimum weather." While some have advocated placing conclusions within table captions—as just demonstrated, most scientists say (and I agree) this is not the purpose of captions.

The ultimate

But if the purpose of a caption is to present information to persons who have not read the text—and all readership studies show that it is—the ultimate way of doing that is in the sort of table and figure captions published in *Science*. That journal's captions sometimes run two or three paragraphs and up to a quarter-page in length. Many scientific societies and publishers are considering whether to require captions of this sort.

An individual author would be unwise to move to such a caption on his or her own. But in the near future it might be just as unwise to write the sort of caption you are used to if the journal you are about to submit your paper to begins requiring much longer and more complete captions than it had in the past.

* * *

Two years after publishing the information presented above, I got to thinking that it might be a good idea to give some specific information on creating and presenting figures. Thus, the following two columns, the first of which was published in August 1994:

The fine art of creating acceptable figures (Part 1)

Until a few years ago, most agricultural colleges employed an artist who, with pen and ink, worked from a scientist's scribbling to create graphs and other figures that were acceptable for publication. Then computers with crude drawing programs and dot matrix printers came in. They made figures with thin, wiggly, and jagged lines and poor-quality type that was not acceptable for publication. Finally (or at least currently) those programs are sophisticated and laser printers make it possible for artists (and occasionally scientists to produce attractive figures.

Of course, scientists who use these programs without the aid of an artist must know what they are doing; otherwise, the figures they produce will be little more than clean and accurate renderings of their scribbling.

This two-part column will present guidelines for artists and scientists to help them produce, if not a work of fine art, at least an acceptable figure. This month I'll explain how a figure should look, next month I'll tell how to prepare figures so they look that way.

Size of final figure

How large should a figure be? The obvious answer is large enough to present the information clearly; but there is more to it than that. Most authors work on a tight budget, and (if they submit their paper to a journal that has page charges) would say a figure should also be small enough to keep the total page charges as inexpensive as possible. Authors also know that when a figure expands from one column to two, its height expands simultaneously, so its total area (and therefore its contribution to the page charge) expands more rapidly than the width of the figure. If two figures have identical proportions and one is exactly double the width of the other, the larger of those figures will take up four times the space of the smaller one on a journal page. Therefore, most authors prefer to present their figures in single columns whenever possible.

The professional editing staff employed by the publisher is aware of this and tries whenever possible to keep figures within one column. But they also must keep figures large enough so that their type is readable.

The lower limit of readability for most eyes is generally 6-point type. Exceptionally legible typefaces can be read as small as 5.5 points and, on rare occasions, even at 5 points. So when the professional editors of ASA–CSSA–SSSA try to reduce a figure to fit into one column they do not let its smallest type go any smaller than 5.5 to 6 points, and try whenever possible not to let it go below 6.5 to 7 points.

Most scientists and artists are familiar with the type measurement term, *points*, so I did not define it here. [In case you are not familiar with it, you are now reading 10-point type.] A related, less familiar type measurement term is the *pica*. A pica is equal to 12 points; 6 picas equal about 1 inch or 2.5 cm. The columns in most journals are 20 picas wide. There is usually a 2-pica alley between columns on a page, so the total width of the type on a journal page is 42 picas. The height of a column of type in most journals can vary, but is generally about 56 to 60 picas.

Wasted horizontal space

After professional editors calculate the reduction percentage that would reduce the type on a figure's X and Y axes to 7 to 8 points, and its smallest type to a minimum of 6 points, they then calculate what the figure's width would be at that reduction percentage. If that width will be 20 picas or less, the figure will easily fit into a single column. If the width will be anything wider than 27 to 30 picas, we have no qualms about putting the figure into a two-column space.

What give us headaches, though, are figures that legibly reduce to a width somewhere between those two numbers. (For some mysterious reason, figures rarely fall into the 23- to 27-pica range: More often than not, the figures reduce to 22 or 23 picas.) This makes them too wide for a single column, but far too narrow for two columns. We don't have many options to remedy 22- to 23-pica-wide figures:

1. We look for horizontal blank spaces in the figure we can eliminate and, thus, allow the figure to be reduced to a single column.
2. We can set larger type and paste it on top of the type already on the figure, which allows us to reduce the figure a few more percentage points so it will fit into one column.
3. We can send the figure back to the authors and ask them to have the type enlarged, the figure redrawn, or both.
4. When deadlines are imminent, we sometimes have no choice but to run the figure in a two-column hole and make the author pay the additional page charge for the wasted space.

Problems with too-wide art can also happen with two-column figures. If a figure reduces to slightly too wide for the 42-pica two-column space, we can use the first three options listed above, plus sometimes we can eliminate redundant type or excess space in figures that are side-by-side components labeled Parts A and B.

If a figure is substantially too wide at a readable type size, we consider running it sideways on a page. Then it can be up to 56 picas wide. Such a figure cannot be more than 40 picas tall at that reduction or the figure and its caption will not fit into the page width of 42 picas. Another occasionally possible alternative is to have the author divide the figure so that part of it fits onto one page and the rest onto a second page.

I recently had a large single-unit illustration with type so small and poor in quality that I could not legibly reduce the figure to a 42-pica width. The figure was too complex for Headquarters to reset its type. I faxed a message (along with a photocopy of how the figure would look reduced to 42 picas) to the author, asking if he could have the figure redrawn with larger type so that it would fit into two columns (and thus occupy about a third of a page). If he could not, I said, we would have to run the figure sideways on the page. This would allow the figure to be 56 picas wide, but also meant that the figure would occupy a full page of space instead of only a third of a page. After reviewing his options, the author said it would be easier and less expensive for him to pay the larger page charge for the full-page figure than it would to redraw the figure.

One of the more common ways an author can waste space is where identical graph boxes are placed side by side. It normally is not necessary to repeat columns of numbers to define their integral marks. A single set of these numbers at the left side generally suffices for both boxes and allows the boxes to be moved more closely together, which might allow the figure to fit into a two-column space, and occasionally into a one-column space and still be legible.

Wasted vertical space

Authors probably waste far more vertical space than they do horizontal space. The most common place where space is wasted is in a two-part figure where Part A is above Part B in a figure drawn on a single piece of paper for one-column reproduction. I have sometimes seen an original with as large a space between those two parts as there was in the figures themselves. Editing out that blank space reduces the size of these figures by as much as 33%. We always edit out this excess space, but the author or artist could have just as easily used less space between the units of the figure as it was being drawn.

If you have a stack of similar elements of a figure with different graphs and points of information, but identical scales, it is usually not necessary to identify the scale on each element. (This is similar to what was described above for horizontal situations.) A single line of type at the bottom of the figure, if the integral marks are the same in all those parts of the figure above it, will usually serve for the entire figure and allow you to draw the individual graphs much more closely together.

* * *

Here's the second of those two papers, from September 1994:

The fine art of creating acceptable figures (Part 2)

Last month I presented information on the way a figure should look in its final form so it will be readable, yet cost the author as little in page charges as possible. This month I describe how to achieve that goal.

Earlier style manuals discouraged use of figures developed on computers. This was simply because the older dot matrix printers and crude programs in use at the time some of those manuals were written could not produce publication-quality art. Today's improved software and laser-printing equipment can create production-quality figures, and the glossy paper made for laser printers is close to the quality of the paper an artist would have used. Even plain paper is generally of high enough quality that, after reduction, it can present publication-quality art.

One question that authors often ask: "Since this figure was produced on a computer, could I send it to you on disk instead of on paper?" The answer is no and yes. Most publishers and typesetter/printers work with hard-copy art because it can be readily reduced and manipulated as necessary. They are sometimes grateful for an electronic backup of that art, though, if the hard-copy version was damaged in transit and the publication deadline was looming. Being able to duplicate the damaged art without taking the time to ask the author to send us another copy can sometimes be a great time saver.

Also, the day is coming [update, is much closer than when these words were first written] when all publishers and typesetter/printers will work with figures within the computer, right along with the text; so it is not a bad idea for you to send your figure on disk to both get yourself in the habit and also to give publishers who want to the opportunity to experiment with it. If you submit a figure on disk, please place it into a separate file from the text on that disk (all figures can be in one file).

Size of original

Most publishers recommend that an original figure be drawn about double its desired final size. You may wonder why you couldn't just draw your figures the final size. First of all, would you like to try and work in that small a space? Of course, this is less of a problem when working on a computer than it would be for an artist with pen and ink.

More importantly, reducing a figure tends to sharpen it. Rough spots in lines tend to smooth out, tiny blemishes in or on the paper vanish, and faintly drawn lines tend to get darker (unless they are too faintly drawn or composed of tiny dots; then they tend to vanish with the small blemishes).

After you draw the figure oversize, don't reduce it yourself before you send it in, because then we cannot take advantage of the benefits described above, particularly that of getting rid of blemishes in the paper.

When you draw the figure about twice the size you want it to appear in the printed journal, make certain the type you use is twice the final size as well. You will want the final size of your smallest type to be at least 6 points, so use nothing smaller than 12-point type on your original figure. (Change all the type sizes mentioned in this paper appropriately if you are planning on a reduction of 40 or 60% or whatever instead of 50%.)

Most publishers recommend that axial numbers and type be no smaller than 8 points. Smaller labels within the graph can go down to a 6 point minimum, and overall labels and titles should be in the 9 to 10 point size range. Which, of course, means that you should use original type that is 12, 16, and 18 to 20 points in order to achieve the desired final size. If you think type this size is too large and gaudy, use your copy machine to reduce it to final size and then see what you think.

Draw your lines thicker than you want them to appear in the final form also; otherwise, after reduction they will be too thin and could vanish or break. Dotted lines can become solid lines after reduction.

If you need to shade some parts of your figure, keep in mind that the spaces between the elements of that shading will be reduced when the figure is reduced. Most patterns built into most computer programs become solid black when reduced to 50% of the original size. Search for patterns or create your own that will not condense to black.

Philosophy of line weights

According to the philosophy stated in our Style Manual, originally proposed by Dr. Dwayne Buxton, former ASA Editor-in-Chief, the most important line on a graph is the derived curve; therefore, it should be drawn the heaviest. Whenever possible, it should not be just a series of straight lines connecting the data points, but an actual curve (with the accompanying equation that defines it printed in the graph).

Next in importance are the individual data points. They should be drawn next heaviest.

Of least importance is the box that contains the material; therefore, it should be drawn the lightest. Encase the figure in a full box, do not just use a single X and Y axis on the bottom and left side. I think this philosophy of line weights makes sense: I recommend it to others and use it myself. It appears that not all scientists agree, for I have seen many a graph that has barely visible curves drawn within extremely heavy boxes.

Another problem and a bibliography

With a good operator, a good drawing program, and a good laser printer, computers can produce high-quality figures; but they may or may not draw a figure the way publishers want it. For instance, several programs (including Harvard Graphics) will automatically put a shadow in your bar graphs, adding a Z axis with no information on it, which makes it difficult to interpret the information. This default setting should be turned off. (See page 237–238 for more information on 3-D graphs.)

Further information on preparing figures is available in the style manuals of most publishers and in a book titled *Illustrating Science*, published in 1988 by what used to be called the Council of Biology Editors, now the Council of Science Editors, 11250 Roger Bacon Drive Suite 8, Reston VA 20190-5202.

* * *

The following is from the February 2000 issue of my newsletter:

Setting tables correctly

There are two electronically correct ways to create tables. You can either use the table function of your word processor or create a table line by line with tab stops. (However you set them, those tables should be separate from the text at the end of your manuscript, not within the text.) I will demonstrate both ways, using as my sample material adapted from a table in the paper: P.F. Pace, et al., 1999, Drought-induced changes in shoot and root growth of young cotton plants, J. Cot. Sci. 3:183–187.

Using table-creation software

The table-creation feature of your word processor is not difficult to learn and can even be fun to use. The one with WordPerfect is awesome. The one with MS Word seems comparable, but more complicated to use. Perhaps it would not be so complicated if I were more familiar with it. I have been able to create any sort of a table I have ever needed with the WordPerfect table maker. Following are the steps I use:

I begin with a simple table grid. It is good to begin with the correct number of horizontal spaces. You can add more horizontal cells, but that can sometimes become tricky and affect the width of each data cell already in your table and, sometimes, the fit of the table on the page.

You do not need to create the precise number of vertical rows, because it is easy to add or delete rows. I normally begin with only about five rows. This allows me at the beginning to set up the columns and headers and get everything correctly spaced horizontally. Then I add rows, usually three to five at a time, as I move down the table.

After I have placed my initial basic (in this case) four-row-deep grid in place, I combine two or three or more cells vertically or horizontally as needed to create the headers for each column in my table. The table-creation software makes it easy to add, combine, or delete cells as needed.

Then I will usually add the type to the proper cells, without regard to how well it fits or whether it is aligned properly. (Note that as I did this, I added the necessary additional rows to the bottom of the table.)

	Treatment	
Plant part †	Drought	Control
T-root length (cm)	24.5 ± 1.4*	18.9 ± 1.2
T-root dry weight (g)	0.260 ± 0.0227	0.260 ± 0.031
S-root length (cm)	52.2 ± 6.7	42.5 ± 4.9
S-root dry weight (g)	0.221 ± 0.030	0.188 ± 0.027

Next, I adjust the size of those cells or type of both to fit. Within those cells, I can set them up for flush left, flush right, centered, decimal, or whatever alignment I need. I can even vary the number of spaces to the right of those decimal points. I can also vary the type face, type size, and type weight (bold, italic, larger, smaller, etc.). In this example I changed the two right-hand columns from flush left to centered.

	Treatment	
Plant part †	Drought	Control
T-root length (cm)	24.5 ± 1.4*	18.9 ± 1.2
T-root dry weight (g)	0.260 ± 0.0227	0.260 ± 0.031
S-root length (cm)	52.2 ± 6.7	42.5 ± 4.9
S-root dry weight (g)	0.221 ± 0.030	0.188 ± 0.027

After I have all the material I want in the table and everything properly lined up, I keep all cells, but delete all the rules on those cells except those I specifically want to retain in the finished table. Finally I add the footnotes and anything else that may be needed.

	Treatment	
Plant part †	Drought	Control
T-root length (cm)	24.5 ± 1.4*	18.9 ± 1.2
T-root dry weight (g)	0.260 ± 0.0227	0.260 ± 0.031
S-root length (cm)	52.2 ± 6.7	42.5 ± 4.9
S-root dry weight (g)	0.221 ± 0.030	0.188 ± 0.027

* Significant at $P = 0.05$
† T-root = taproot; S-root = secondary root

In more specialized situations I can also add background shading, or anything else necessary to complete the table. I have even added formulae to tables and turned them into something approximating a spread sheet.

Using tab stops

Or, you can set up a table the way you used to with a typewriter. If you do, use only one tab stop between each entry in each line. You can do this by using the various kinds of tabs available, instead of using only the "hard left" tabs you are likely used to for all or even individual lines in your table.

Word processing software places a wonderful array of tab stops at your disposal. WordPerfect, for instance, has eight different tabs, include left, right, and decimal. In the space where you have keyboarded your table, delete all the tab stops and then insert any combination of the correct stops that you need to align your columns. For those header lines at the top and those data lines that do not contain as many items as other data lines in your table, highlight those lines (one at a time or in sections) and change the *stops* for those lines. The important thing to remember is to have only one tab between each item in each line of your table, even if it means that each line in a table needs to have its own set of tab stops.

Using the same information as before, here is the process:

First, set the material with a single tab between each item:

Treatment
Plant part † Drought Control
T-root length (cm) 24.5 ± 1.4* 18.9 ± 1.2
T-root dry weight (g) 0.260 ± 0.0227 0.260 ± 0.031
S-root length (cm) 52.2 ± 6.7 42.5 ± 4.9
S-root dry weight (g) 0.221 ± 0.030 0.188 ± 0.027

Second, delete all tab stops at the top of the table (do not panic when you see what happens). This demonstration was created with WordPerfect, which stacks everything. MSWord keeps the material in its rows.

Treatment
Plant part †
Drought
Control
T-root length (cm)
24.5 ± 1.4*
18.9 ± 1.2
T-root dry weight (g)
0.260 ± 0.0227
0.260 ± 0.031
S-root length (cm)
52.2 ± 6.7
42.5 ± 4.9
S-root dry weight (g)
0.221 ± 0.030
0.188 ± 0.027

Third, set the sort of tab stops you need; in this case, only two centered ones:

	Treatment	
Plant part †	Drought	Control
T-root length (cm)	24.5 ± 1.4*	18.9 ± 1.2
T-root dry weight (g)	0.260 ± 0.0227	0.260 ± 0.031
S-root length (cm)	52.2 ± 6.7	42.5 ± 4.9
S-root dry weight (g)	0.221 ± 0.030	0.188 ± 0.027

Fourth, highlight the first line only, and change its tab setting:

	Treatment	
Plant part †	Drought	Control
T-root length (cm)	24.5 ± 1.4*	18.9 ± 1.2
T-root dry weight (g)	0.260 ± 0.0227	0.260 ± 0.031
S-root length (cm)	52.2 ± 6.7	42.5 ± 4.9
S-root dry weight (g)	0.221 ± 0.030	0.188 ± 0.027

Finally, add footnotes, rules, etc:

* Significant at $P = 0.05$
† T-root = taproot; S-root = secondary root

Chapter 11

References

The column you are about to read carries a lot of significance for me. One day in early 1992, as I was in the process of editing the third or fourth paper with the same problem, I shouted in frustration to nobody in particular, "Doesn't anybody know how to alphabetize a reference list!?" I immediately converted my frustration into action by turning to my typewriter (that's how long ago it was) and writing a little essay titled, "How to alphabetize a reference list." For some reason, just before I handed it to the editor of what was then called *Agronomy News*, I added the line "Fine-tuning your writing" above that main title. Eight years and 92 columns later I retired from ASA–CSSA–SSSA and founded a writing-editing-consulting firm by the same name. Here is where it all began, a column published in March 1992:

How to alphabetize a reference list

Headquarters receives too many papers with reference sections that are not alphabetized correctly. While staff editors can correct them, it is not always easy, particularly for papers that have numbered references. Also, too many authors single space their reference sections, which makes it difficult to make the needed corrections.

This paper, first in a series of writing tips, will outline the steps

---

for correct alphabetization. Basically, reference lists are alphabetized by author name. Chronological order is not considered except when names are identical.

Use the following sequence to determine the order of the names in a reference list:

1. By the last name of the first author.
2. By the first initials of the first author:
   Jones, A.L.
   Jones, C.B.
3. By the second initials of the first author:
   Jones, A.A.
   Jones, A.B.
4. All single-author entries precede multiple-author entries for that senior author. List them chronologically and, when there are two or more in the same year, alphabetically by title. Each same-year entry is given a lowercase letter to designate the order in text citation and reference list: Jones (1991a).
5. By the last name of the second author:
   Jones, G.W., and D.A. Haliburton
   Jones, G.W., and C.G. Jamieson
   Jones, G.W., and A.C. Manocotti
6. By the first initial of the second author:
   Jones, G.W., and D.G. Haliburton
   Jones, G.W., and F.A. Haliburton
7. By the second initial of the second author:
   Jones, G.W., and D.A. Haliburton
   Jones, G.W., and D.B. Haliburton
8. Repeat steps 5, 6, and 7 as needed for the third, fourth, fifth, and sixth authors. (A reference to any paper that has more than six authors should use "et al." after the sixth name. The only exception to this rule applies to Crop Registration papers in *Crop Science*; all authors should be listed in references to these papers.)
9. When authors are the same, list in chronological order:
   Jones, G.W., and D.A. Haliburton. 1989.
   Jones, G.W., and D.A. Haliburton. 1990.

   As in step 4, references that are identical even to the year are listed in alphabetical order by the title and given a lowercase letter after the year in text citations and in the reference list

10. In any case where the text citation would be identical, even if the reference is not, a lowercase letter must be appended to the year. In the following example, without the letters, both publications would be cited as "Jones et al. (1991) within the text:

    Jones, A.B., S. Muldowney, S. Fisher, and I.M. Hungry. 1991a.

    Jones, A.B., J.P. Zenger, and T.A. Balderdash. 1991b.

* * *

What you may or may not have noticed in that column is my penchant for having fun. While providing information I also slip in humorous, esoteric, and trivial information. Shirley Muldowney was one of the first successful woman drag-racing drivers, J.P. Zenger is a famous person in the history of American journalism, manocotti is a type of pasta, and balderdash speaks for himself. (The S. Fisher was a change by the editor from S. Fetchit. Stepin Fetchit was the first African American to become successful in motion pictures. Later generations would see his success as a mixed bag because the roles he played were quite stereotypic—which may be the reason why the editor changed the name.) And, for those for whom the above was too esoteric, I also used the name I.M. Hungry, which I figured everybody would understand as a joke.

Getting more serious, in the final volume of the original columns I gave the reasons why an author should be careful in preparing his or her citations and reference lists. From December 1998:

Who cares about details? Readers and you

Sometimes I'm embarrassed to ask the questions I need to ask in a query letter. So often those questions are about a seemingly trivial matter. I'm sure at least some authors' first reactions will be, "What difference does it make!?"

Like, for instance, you have a citation of "Smith and Jones, 1998" in your text and in your reference section you have a paper by a Smith and Jones published in 1989, and I ask you "Which is correct, or are these two separate papers?"

Or perhaps in your reference section you list authors of a paper as "Jones et al." with no initials and no other names, and I ask for more specifics.

I'm sure at least some authors think little errors such as these make no difference in the presentation of science in their paper. These errors *do* make a difference, though. They can, in fact, make a big difference to the reader who wants to try to duplicate your research, a key element of science.

If readers or librarians search all available sources they will likely find several papers, books, or what-have-you published by persons named Smith and Jones in 1989 and several more by persons with those names that were published in 1998. Which is the particular one (or two) that you referred to in your paper? It could be difficult for the reader or librarian to find out.

And if the particular paper the reader sought contained information essential to verifying the results of your experiment, the researcher– reader faces a quandary: "Do I contact the author for a clarification, or did the author, perhaps, fabricate the reference and would it be better to let sleeping dogs lie?" (You see, such an error is not nearly as insignificant as you might like to think.)

As to the second problem cited above (that of several people with the same name), there are more people than you might suspect with similar names. Jones and Smith are far from the only names that fall into this category. For instance, it not too long ago that three people in Illinois named Mulvaney were publishing papers. They happened to be father, son, and daughter in law. We have several husband-and-wife pairs, parent-and-offspring pairs, sibling pairs, as well as many others who are not related, but happen to share the same surname and sometimes even the same initials or given names. For example, I know, of at least three Gary Petersons (not related, so far as I know) who regularly publish in our journals.

Time to change hats

There is no question that your role as you are writing the first draft(s) of your paper is that of the author of that paper. As you move from the writing to the revision stage(s), though, I suggest that you set aside your author's hat and don the hat of the reader, or the reader's reference librarian.

It takes a little effort on the part of the writer to make certain the little details are complete, but it takes a far greater effort on the part of readers and reference librarians if you do not, because readers and librarians do not have the benefit of your knowledge.

I must confess to a personal bias against people close to the information who do not provide that needed information. This bias began on the first job I held after college. The editor of that newspaper was an excellent photographer himself and staffed his paper with other photographers at least as good as he. Every photographer on the staff had won at least one national award. As you might suspect from what I've just written, the paper's photographers were prima donnas and the copy editors and reporters were not.

It was not unusual for that editor to drop a photograph onto my desk and say to me, "Write a caption about this."

"Who is this a picture of?" I would ask.

"That's for you to find out."

"Where was it taken?"

"Northwestern Illinois."

"That's a fairly large area. Can you be more specific?"

"No."

"When was it taken?"

"Yesterday."

"Can you even tell me the time of day it was taken?"

"No."

Being new on the staff of that paper, I was the lowest of the low on the totem pole, so I didn't dare ask the obvious question: "The damn photographer was there, why couldn't he either write the caption himself or at least gather enough information to give me a clue as to where to start looking for the information?"

The first reader

Editors consider themselves the first readers of a paper and therefore represent all subsequent readers. Editors try to ask the questions readers would ask. We try to dig out and correct inconsistencies that would confuse subsequent readers. The reason for this is simple, if you think for even a few seconds about the frustrations you may have faced reading someone else's paper that was published in a journal whose editors are not as careful as those employed by or volunteering their expertise to your three Societies.

So, if some editor asks you what you consider to be a dumb question, realize that it is not a dumb question at all, but more likely a question that a reader would logically ask. You can either answer the editor once, or refuse to answer the editor and find yourself answering the same question hundreds of times from many readers after the paper is published.

Or, better than waiting for the editor to ask the question, put yourself in the mind set of a reader as you are revising your paper. Anticipate those questions yourself and answer them before an editor sees them.

Not just the names

One logical place to begin is with your references, because they generally prompt the most questions. I began this column with questions that can come up with incomplete or confusing names in references. There are a host of other things to also think about as you construct a reference list. If you are citing a chapter in a book, for instance, the Societies require that you give the name of the chapter author(s), the title of the chapter, the name of the book's editor(s), the correct title of the book, when it was published, and by which publisher.

Proceedings are perhaps the trickiest citations of all. Usually the publication of the proceedings comes out a year or two after the event itself, and the address and name of the publisher may be different from the original sponsors and the place where the conference was held. (A conference last year in Nairobi may be published next year in London.) That's why proceedings references require not only the title, date, and location of the event itself, but also the page numbers of the particular cited paper, the name(s) of that paper's author(s), the name(s) of the editor(s), the name of the publication and the year it was published, and the name and address of the publisher.

I'll grant you that it is a nuisance to gather all that information and a nuisance to put it into the paper. It is even a greater nuisance if you do not supply that information in the first place and we have to ask you for it later. But a still greater nuisance would be the nuisance to the reader(s)—or reference librarian(s)—if you fail to provide the information they need to find the publication you cite.

So proudly wear your author's hat as you write the best manuscript you possibly can. Before you submit the manuscript, though, exchange that author's hat for a reader's hat and do your best to make your manuscript reader friendly. If you don't, you will hear from people who are paid to wear an editor's hat.

(And thanks to Jon Bartels, former ASA–CSSA–SSSA Headquarters managing editor for books, monographs, and special publications, for suggesting this topic.)

Chapter 12

Statistical and Scientific Presentation

I'm the first to admit that math and statistics are not my strong points. But I'm also quick to admit that I have some strong biases about the way statistics are presented in some papers. Perhaps it stems back to the old adage claimed to have been first spoken by so many persons that it would take nearly a book to list them all. At any rate, the adage goes something like this, "There are three kinds of lies: Lies, damn lies, and statistics."

Some other aspects of scientific nomenclature have also caused difficulties through the years and will be dealt with in this chapter as well.

I begin, though, with statistics, and confess that I have tried to keep authors honest in their use of statistics, with greater or lesser success through the years. At any rate, here was my first foray into the subject, co-authored with Gary A. Peterson of Colorado State University, and published as my third column in May 1992:

Statistical significance: How much is too much?

Present-day computer programs allow authors to be precise in the levels of statistical significance they report. We have seen, for instance in the text of a single page, the following levels: 0.01, 0.001, 0.002, 0.004, 0.018, 0.038, 0.042, and 0.047. Just because authors can do this, does it follow that they should?

For the majority of the papers submitted to our journals the author could simply establish a particular significance (at the beginning of the paper or perhaps at the beginning of its "Results and Discussion" section) and then simply say from that point forward, "this result was significant" or "this result was not significant."

Within the text of a paper the author should refer to "significant" or "not significant" based on an alpha level of his or her choice. It simplifies material for the reader and still allows authors to choose what they consider a real difference. The actual probability level still could be reported in the tables. Consider, for instance, a paper that presents one finding as "significant" at 0.05 and another finding as "highly significant" at 0.01. It would be more appropriate to just indicate that the difference was significant, with the specifics given in the tables.

Also frequently seen in journal papers: "this was significant at $P =$ 0.05, this other thing was significant at $P = 0.01$, this was at $P = 0.001$, and this was not significant." While most would argue that the lowest level would apply to the negative, this is not clear when multiple levels have been referred to in previous sentences. The level an author sets for a particular paper would be determined by the nature of the work. For some papers $P = 0.1$ (or even 0.25) would be sufficient. For other papers, 0.001 might not be precise enough.

There are probably a few instances where an author could legitimately cite any probability level at various points in the text. We believe those exceptions are just that—exceptions, and not the rule.

Researchers, as you are setting up your work and determining your results, take a realistic look at your work and determine a level of significance you need to properly assess your results.

Authors, as you begin to write up your research results, determine how many of the minute details you have gathered actually belong in the paper and which beyond that only clutter the results.

Editors and reviewers, use a critical eye as you review and assess the significance levels used in papers. Is there a need to discuss as many different levels of significance as the authors do? Do the multiple levels contribute to precision or confusion.

And when you are reporting your results in tables, be aware of both your format and the clarity of your footnotes. Often it is difficult for an uninitiated person to tell what is being compared. Using a footnote format such as the following (based on one developed by former Headquarters staff editor Gundega Korsts) will eliminate ambiguity:

"Within ____________, means followed by the same(different) uppercase(lowercase) letter are(are not) different at the ________ probability level." Explanation: This format states (i) what is being compared; rows, columns, traits, years, combinations, etc.; (ii) whether to look for the same or a different letter; and (iii) whether it is significance or nonsignificance that is being demonstrated.

* * *

I revisited the subject in November 1995 from a slightly different perspective, with the help of the same Gary Peterson and also Dr. Jerry Hatfield (director of the National Soil Tilth Lab):

How to review statistics in a paper

The most common complaint I hear is, "Our journals are too rigid." Unfortunately, there is some truth to these complaints; but I believe reviewers, and not editorial boards, are the main source of this rigidity.

I don't know who first said or wrote it, but I've seen the following posted in many editorial and publishing offices: "The strongest drive is not love or hate, it is one's need to change another's copy."

Scientists follow this dictum as frequently as anyone else, and each has an opinion of how a sentence should be structured—use passive voice, don't use passive voice; always place a comma here, never place a comma here, etc. Authors, of course, would rather follow their own grammatical whims than be forced to follow someone else's.

Complicating this picture is the fact that scientists review papers for several journals and may not have an intimate knowledge of ASA–CSSA–SSSA rules. For example, they may ask an author to follow another journal's rules and make unnecessary or even improper changes in their manuscripts. Our editorial board members try to buffer reviewers' comments of this sort, but are not always perfect.

Authors who were forced to make unnecessary or improper changes to get their paper accepted resent it when a member of your professional editing staff in Madison asks them to make legitimate changes. More than a few authors wonder if their paper will *ever* get published.

I apologize if reviewers were unfair with your paper, but suggest you don't take revenge by chopping apart the next paper *you* review (unless the paper deserves chopping apart). If the only problem with a paper is that its author does not write the same way you do, consider that a quirk of the author, not a scientific flaw. If the science is correct and the sentences are understandable, try to let the personality of the author show through. It is the author, after all, whose name is in the byline, who will pay the page charges, and who must defend the integrity of the work.

Statistics present a deeper problem than word usage because it involves scientific integrity.

Author prerogatives

An author may *know* that, compared with more common methods, the statistical method chosen allows a much more rigorous test of the hypothesis associated with the objectives of the particular research project. A reviewer may *believe* some other statistical method is the *only* one to use in *any* situation. The italicized words speak for themselves.

An author chooses a particular significance level based on the risk he or she is willing to take regarding rejection or acceptance of a hypothesis. While one researcher may be quite content with a level of $P = 0.1$ for his work, another will say $P = 0.0001$ is not precise enough for her needs. Unless the level chosen was not appropriate for the subject under study, a reviewer should not recommend rejection of a paper simply because the author is more likely than the reviewer to accept a Type 1 error.

Information updating

You would naturally resent it if a person who had not taken a class or worked in your specialty for several years tried to tell you how to do your work. After all, that person (i) is no longer current in the subject and (ii) your way is acceptable, and probably better, than his or her recommendation. The same is true with statistics; like any other branch of science, its practitioners continue to develop new techniques. If you haven't taken a statistics class in the past five or 10 years, you will be no more current in that specialty than others would be in your field if they had not worked in it for a comparable time.

If the author of a paper you are reviewing used an unfamiliar statistical method, don't automatically discount it. First ask yourself if the procedure is valid. If it is, it is probably all right. If you cannot tell the validity for certain, ask for a second opinion from someone familiar with current techniques. If you don't have time for that, review what you know in the paper and return it to the editorial board with an indication that you are not familiar enough with the statistical method to comment on it.

Your Societies are trying to keep their editorial boards as current as possible in statistics so they can properly review statistics in the papers and, when necessary, be able to buffer reviewers' comments and suggestions. At the 1994 annual meetings in Seattle, editors and technical editors of all our journals attended a seminar on the strengths and weaknesses of the newest statistical techniques. A program for all members of all editorial boards was scheduled for the 1995 meetings in St. Louis. These statistical training sessions for our editorial boards will probably continue at future annual meetings as well.

If you want to refresh your own statistical knowledge, consider taking a class or attending one of the pre-meeting seminars on geostatistics that your three Societies often hold before their annual meetings.

* * *

As a brief update, various sorts of statistical training have been held off and on since 1995 for members of ASA–CSSA–SSSA editorial boards. Whether they have received enough training is a matter of opinion. No doubt some authors will believe that nobody on any editorial board will have had enough training to review their papers. My most recent foray into this subject was in my newsletter in February 2000:

When differences aren't statistical

I get awfully tired of reading about supposed differences in statistical significance that are probably not different. Take this quote from a recent paper [details deliberately obscured]: "Differences in [measured] levels between treatments were significant ($P = 0.05$) only on [one date] (Table 2). However, on [another date], differences were significant at $P = 0.06$."

The wonderful statistical packages available nowadays allow a person to write something like this. But what are authors actually saying when they do? Statisticians I have talked to tell me that the only use for significance is to establish a particular level at the beginning of a study—be it $P = 0.05$ or 0.01 or whatever. Then whatever differences you find significant at that level are "significant" and those you do not are "not significant," period.

Some people, possibly like the author of the passage quoted above, take it one more step. That is to establish a level, say $P = 0.01$, and says from then on that any finding smaller than that was significant, and to give the numbers of those larger than that to show how close it might have come to the established level of significance. "It was not significant (P = 0.07)" vs. "(P = 0.1)."

Still others view it in a slightly different way. Forget, they say, about establishing a level and simply report all the probability levels and let the reader decide what is important. I think that is an abdication of one of the roles of a scientist, that is to interpret the work he or she has performed.

But a larger issue is involved here. Some statisticians fear that at least some scientists are trying to rationalize along the lines of, "well it wasn't what we wanted as a probability level but it was really close so we'll think about it as being significant." This comes close to crossing the line between abdicating responsibility and participating in fraud.

I think at least some scientists in some cases—for reasons only they know—are trying to detect a difference within the error of measurement.

All the above reminds me of a paper I read once by an undergraduate of a study he had conducted by himself. His paper explained the setup of his experiment, how it was conducted, how the results were gathered, and reported those results, which he admitted were not statistically different from the control. Then he concluded with this gem: "But farmers aren't interested in statistical significance, they are interested in profits. And I have just demonstrated that my technique will increase their profits."

I sure hope he either never made it into the profession, or that he learned something about statistics before he went much further.

* * *

Another area of nomenclature that can cause problems is molarity vs. normality. Increasingly, journal style calls for the former instead of the latter, which used to be far more common in journals. This column on that subject was published in August 1996:

Use moles to express acids

This month's column will be a little different from normal. The "Nomenclature" segment of my April column [Chapter 15 of this book] attracted a rebuttal from a Society member. This column will reprint much of that letter, plus a reply to it from an authority on the subject.

I will not reveal the name of the author of the original letter so as not to embarrass anyone, but want to help other authors who may also misunderstand use of the molar term. First, pertinent parts of the initial letter:

"Converting normal values to molar violates standard chemical practices. While the concentrations of ions and molecules should be expressed as molar, the only unambiguous way to express the concentration of acids and bases is with normality. By convention, we know that a 1 *M* solution of NaCl contains one mole of Na^+ and one mole of Cl^- per liter and a 1 *M* solution of $CaCl_2$ contains 1 *M* of Ca and 2 *M* of Cl per liter.

"With acids and bases, the number of moles of H^+ or OH^- are normally of interest and not the number of moles of the accompanying ion. A 1 *N* solution contains one mole of H^+ or OH^-. For NaOH, a 1 *M* solution equals a 1 *N* solution. For H_2SO_4, the situation is ambiguous. Because the H^+ is the ion of interest, does a 1 *M* solution contain one or two moles of H^+? By using normality, a reader knows for sure that a 1 *N* solution contains 1 *M* of H^+ or OH^- per liter. Rather than discouraging the use of normality, we should adopt a policy of requiring authors to use normality when referring to acids and bases."

Investigation

This question involved scientific, not grammatical style. The style stated in my April column was directly out of the Societies' *Publications Handbook and Style Manual*, so the first thing I needed to determine was whether there was an error in that policy. If so, it would have to be changed, which would require persons with more authority than me.

You might be interested to know that the Boards of Directors of the Societies appointed two scientists as S.I. usage consultants to the editorial boards of our publications and to the professional editing staff at Headquarters. One of these consultants has expertise in soils, the other in crops. These two people do not have the time to field S.I. style and usage questions from all members, so I will not identify them here. Trust me (and the people who appointed them) that they are competent individuals.

I forwarded the complete text of the above letter to both of these consultants with the following note: "Please help me out on this one; have we been doing something wrong? [This letter] was obviously prompted by my column in the April *Agronomy News*, but that column merely reported on what we have been doing for a long time."

Response

Both of the consultants responded similarly. The reply from one with expertise in soils was somewhat more detailed than the response from the other, so I will quote the relevant parts from only that letter:

"No, we have not been doing anything wrong with regard to the molar vs. normality question . . . . I think the concern raised stems from the incomplete way our authors are using the mole term. Use of the mole term to designate an amount of substance requires a two-part application, whereas too many times authors only fulfill the first part. The two parts come from the definition of mole adopted in 1971 by the 14th International Committee on Weights and Measures.

"Part one says, 'The mole is the amount of substance of a system which contain as many elementary entities as there are atoms in 0.012 kg of ^{12}C.' This is conventionally known as Avogadro's number and assigned the value of 6.02×10^{23}. Our authors are universally familiar with this concept and it should not be cause for concern.

"Part two says, 'When the mole is used, the elementary entities *must* be specified and may be atoms, molecules, ions, electrons, other particles, or specified groups of such particles.'

"Thus, the use of the term *mole* requires reporting both the quantity and the nature of the material in question. Too many times authors fail to explicitly identify the nature of the material being described, in essence leaving it open to the readers' interpretation. I think this is exactly the case as described by [the correspondent]. If the concentration in question is reported as X moles (H^+) then there is no confusion, regardless of whether the source of the H ions is HCl or H_2SO_4. Solutions containing ions of mixed valence should be described on a molar basis for each ion.

"I do not think a new policy is needed, just a tighter requirement for authors to utilize both parts of the mole's definition. The last sentence on page 39 of the *Publications Handbook and Style Manual* (1988) says, "When the unit mole is used, the entities must be specified." [The comparable statement in the 1998 edition of that book reads, "With this definition, the elementary entities must be specified and may be atoms, molecules, ions, electrons, other particles, or specified groups of such particles. The substance may be a mixture, such as air."]

What this means

I know from experience that molarity vs. normality is a fairly common problem. I have had to ask more authors than I can count to convert normality to molar units. It is clear from those cases that (i) not all of our authors are aware of the requirement for molar units in papers, and (ii) not all members of our editorial boards have been making the needed conversions or asking authors to make them.

In fact, I wrote the April column in an attempt to clarify this matter. Obviously it did not, or it would not have prompted the letter that became the basis for this column. The essence of this column was distributed to members of all ASA–CSSA–SSSA editorial boards.

Chapter 13

Passive voice

Nobody who is familiar with my writings through the years will be surprised that I included a chapter on passive voice. I have written and spoken much against this form of writing through the years, some of which I was unable to publish until I created my own newsletter. Those of you who are not newsletter subscribers will see some of that material here for the first time. And all of you will also see some additional information supporting it that you likely have not seen before.

My first piece was published by ASA–CSSA–SSSA in June 1995:

Here's what's wrong with passive voice

Most English sentences have a similar construction; *subject-verb-object*. We've read these declarative sentences for years, beginning with, "John runs fast," and advancing to "The American Society of Agronomy, Crop Science Society of America, and Soil Science Society of America will hold their annual meetings this year in St. Louis, MO."

Readers have been conditioned by years of reading these declarative sentences to look for the subject at or near the beginning, the verb closely following, and the explanation next, with the most important word or words at the end. This format is common not only because of author habit, but also for the sound psychological reason that readers most readily remember the first and last parts of anything. (Many more remember the beginning and ending of Lincoln's Gettysburg Address, for example, than remember the entire speech.)

This quirk of memory works the same way with parts of individual sentences, whose beginnings and ends are sometimes called the sentence's *stress points.* If you want readers to quickly understand and remember what you write, put the most important words of each sentence at the beginning and end.

You cannot do this in passive voice because passive voice scatters the important words and places the weakest words at the stress points. When readers don't find important words at these stress points they must reread the sentence (at least in their mind) to find its meaning. Readers who have difficulty understanding a sentence's meaning usually blame themselves for not concentrating. But more reader concentration will not solve problems caused by illogically constructed sentences.

Simple example

To demonstrate, let's take a simple sentence written in passive voice: "His father and grandfather were engaged in hybrid seed corn and livestock production."

Many who read this sentence say, "I see nothing wrong with that; I can understand it." It's true, you *can* understand it, but not as quickly as you could if the author had eliminated two major flaws.

Before I tell you what the sentence's flaws are, read this corrected version and see if you can spot them yourself: "His father and grandfather produced hybrid seed corn and livestock." Most readers, even if they can't explain why, say the second version seems, maybe even *feels*, better.

The incorrect sentence begins well; the subject is clear and positioned where a reader expects to find it: "His father and grandfather."

The sentence's first major flaw is its verb, "were engaged in;" it's in the proper place, but vague and weak.

After reading the subject and verb, the reader naturally wants to know what the father and grandfather "were engaged in." A glance at the end of the sentence tells the reader it was "production." Production of what? To find *that* out, (the second major flaw) the reader must reread the sentence, or at least reconstruct it in his or her mind, and learn it was hybrid seed corn and livestock.

The corrected version of the sentence replaces the weak passive verb, *were engaged in,* with the strong active verb, *produced*; and—at the end of the sentence where the reader expects to find it—clearly tells what these gentlemen produced. Nothing slows or hinders reader interpretation.

More difficult example

If you had to slow down, even a little, to interpret this simple passive-voice sentence, how much would you have to slow down to understand a really complex sentence? Let's use one from an actual paper to find out: "The solubility method is a widely used approach for obtaining thermodynamic data for soil components due to its simplicity and high sensitivity."

A reader using the standard *subject-verb-object* stress points would first interpret this sentence to mean: "The solubility method is a widely used approach due to its simplicity and high sensitivity."

While accurate, I believe the author wants us to take more than that from this sentence. I'm certain the author wants *thermodynamic data for soil components* in there somewhere. Let's put it in and see if we can makes sense with it: "A widely used approach for obtaining thermodynamic data for soil components is the solubility method, due to its simplicity and high sensitivity."

The opening is now so long we cannot tell whether the subject is supposed to be the *approach used*, or *obtaining the data themselves*.

Let's guess that the author wants to stress the latter. While we're at it, let's also simplify *widely used approach*: "To measure thermodynamic data for soil components, many scientists use the solubility method because it is simple and highly sensitive."

The beginning and middle are better now, but the ending is vague. Without the final phrase it could mean, "To measure thermodynamic data for soil components, many scientists use the solubility method."

If the author wants that final phrase left in, though, the sentence might mean one of two things. The first: "To measure thermodynamic data for soil components, many scientists use the simple and highly sensitive solubility method." The second: "Many scientists use the simple and highly sensitive solubility method to measure thermodynamic data for soil components."

In other words, after all this work, we still cannot tell how the author wants us to interpret this sentence. The sentence's main problem was lack of clear presentation, not passive voice. Quite likely, the passive voice construction masked the sentence's main problem even from its author.

Benefit yourself, too

You might think from what I've written thus far that the reader is the main person who benefits when sentences are written in active instead of passive voice. Actually, authors benefit too, in at least two ways.

The first is intellectually. I touched on it in the example just given; the passive voice construction made it difficult for the authors to see the error of logic in their sentence. Had they written in active voice, the authors probably would have corrected the ambiguity before they submitted the paper.

Second is monetarily. I shortened the sentence about the father and grandfather from 13 to 10 words (23%). The second example, if one of my versions proves correct, shortened from 22 to 17 words (18%).

I have frequently eliminated 30% and sometimes as many as 40% or more of the words in sentences by restructuring and converting from passive to active voice. If an author who writes in passive voice were to convert most of those sentences to active voice, the total length of the paper could be reduced at least 5%, and perhaps as much as 10%, or even 20% or more. This would lead to a similar reduction in page charges, if the paper was published in a journal that charged them.

Case for passive voice

I do not want to leave you thinking you should never use passive voice. I can think of at least four legitimate uses for it.

One is to soften an uncomfortable subject. Consider this sentence: "We killed the test animals." This seems like an unnecessarily harsh way of presenting a fact, particularly in view of some people's feelings and actions toward use of animals in research.

An author who tries to keep a sentence like this in active voice might first soften the verb to a weaker one, such as: "We painlessly killed the test animals" or "We euthanized the test animals." These do not soften the sentence as much as one might suppose, though, because the key words fall in the stress positions in the original sentence and in both attempts to soften it. The reader subconsciously reads all three in exactly the same way: "WE KILLED ANIMALS!"

With passive voice, though, a person could write, "The test animals were euthanized." Only weak words fall at the stress positions now. The reader would interpret this as: "Animals were euthanized," which, as far as the reader's subconscious mind is concerned, may have happened hundreds of years ago in another part of the world.

A second use for passive voice is to hide credit or blame. "Were euthanized," in the above sentence eliminates the need to say precisely who did the nasty deed. Authors can use passive voice when they describe some part of a project they didn't conduct themselves: "After *we collected* the soil samples *they were analyzed* in a licensed lab." If the authors did none of the work, they should write, "The soil samples *were collected and analyzed* in a licensed lab," or perhaps, "The soil samples *were collected and transported* to a licensed lab where they *were analyzed*." If the authors actually did the work, though, it's all right for them to admit it.

A third acceptable use of passive voice is to inject variation into writing. A long string of "We did this. We did that. We did something else." can get monotonous. An occasional sentence written in passive voice gives your reader a respite: "We did this. Then that was done. Then we did something else."

The fourth use was demonstrated [passive voice used here for variation] in the final paragraph of the previous section: I used passive voice in two sentences to keep them gender neutral: "If an author who writes in passive voice were to convert most of those sentences to active voice, the total length of the paper could be reduced at least 5%, and perhaps as much as 10%, or even 20% or more. This would, of course, lead to a similar reduction in that author's page charges . . . ."

If a sentence in your paper does not present one of these four needs, though, I suggest you use active voice. It will (i) reduce the chance readers will misinterpret your meaning, (ii) help you see your material more clearly so you can eliminate errors of logic, and (iii) reduce your page charges by shortening your paper's overall length.

(I took all the examples in this column from papers I've handled. Most of the thoughts are my own, but I credit some of the logic to "The Science of Scientific Writing," by George D. Gopen and Judith A. Swan, published in the November–December 1990 issue (Vol. 78) of *American Scientist*. Dr. Gopen has expanded this paper into a book. I have read its manuscript, but the book itself will not be published for a couple more years. I'll give you its title when I let you know the book is finally available. Any good writer and anyone who wants to become a good writer will want—perhaps *need* is the better word—this book.)

* * *

This second column, giving additional information on the same subject, was first published in February 1997:

Take another look at passive voice

Professor Steven J. Knapp writes: "Would you mind preparing another column on active vs. passive voice, this time concentrating on the first person singular vs. third person passive voices? I have had a tremendous number of problems with getting certain manuscripts published during the last five or so years because I wrote them in the first person and used an active voice. Several *Crop Science* editors are among those who have criticized my use of the first person. Readers and referees seem to object to 'we did this or we did that,' and ask us to state 'this was done or that was done.' This seems foolish to me, and decreases the quality of the writing. The *CBE Style Manual* states: 'Avoid the passive of modesty, a device of writers who shun the first person singular.'"

Why is passive voice so prevalent?

I suppose one could make a deeply philosophical argument for passive voice. A few scientists, I'm sure, find it difficult to write, "I discovered this," because they realize that virtually every finding was built on earlier findings which, in turn, were built on still earlier findings. While admirable in the purest sense, I can't help thinking there is too much false modesty involved when someone refuses to use active voice.

Still, the roots of passive voice run deep. I think the main reason why passive voice is so prevalent in scientific writing is the simple fact that everyone does it. If all the scientific papers you read in high school were in passive voice; if all the graduate students you knew while you were an undergrad wrote in passive voice; if your major professors while you were a graduate student wrote in passive voice, then you would naturally come to believe that passive voice is the proper way to write.

A phrase such as "we did this" sounds to some scientists more like something a person would read in a grade school primer than in a *scientific journal*. I mean, after all, a *scientist* would not write, "We did it because we wanted to figure out how it works." A *scientist* would write, "The research was conducted to ascertain the function of the system."

Passive voice discouraged

Dr. Knapp quoted the *CBE Style Manual* accurately. Although some might find it difficult to believe, the Societies also have tried to discourage too much passive voice. A few years ago a committee drafted a report titled, "Become a more successful author," which is still published periodically in all of our journals. One of the statements in it makes a strong case for active voice. That statement drew the following comment during the review process, "Won't we have to change the by-laws before we can publish this?" I don't know if the reviewer was serious or joking, but there is no law that says scientific papers must be written in passive voice.

Origins

Where did passive voice begin? I don't know. It has been around for a long time, though. For example, I looked at some papers in the 1910 issue of what is now called *Agronomy Journal* and found statements so vaguely passive I could not even tell if a research project had been undertaken. Consider the following paragraph. Is it a description of a project or an historical summary?

"On the experiment field at Cornell University are a number of experiment plats having an area of .01 acre. The dimensions of these plats are 43.5 × 10 ft. As it is customary here to plant corn in hills 3 × 3 ft. each plat contains 39 hills of corn. Each plat is duplicated or triplicated, but in spite of that the limited number of plants on a plat might lead to an error if certain individuals on one plat differed greatly in productiveness from those on other plats, even when all are of the same variety. In order to decrease the probability of variation in individual productiveness the plan of planting all of the plants in an experiment with kernels from the same ears has been adopted. If, for instance, there are thirty plats in the experiment, thirty small sacs are placed in a row and a kernel from the same ear is placed in each of the sacs. This is repeated until the number of kernels required to plant a plat is deposited in each sac. The number of ears used is sufficiently large to prevent the possibility of any injurious result from close breeding." (From "A test of planting plats with the same ears of corn to secure greater uniformity in yield." T. Lyttleton Lyon, *Proceedings of the American Society of Agronomy*. 2:35)

I presume Lyon was writing of his research project, but it could just as well have been a description of the way things had been done at Cornell for the previous 50 years.

After reading that passage (and others like it in old journals as I was gathering material for this column) I couldn't help thinking that progress has been made in the past 86 years. Not nearly enough progress, but at least the tide is turning.

Uses for passive voice

You need not completely avoid passive voice. I outlined in the June 1995 column four legitimate uses for passive voice, and have since realized a fifth: If you are writing a person's biography, you may write, "He was born on 7 Dec. 1941," you do not need to write, "His mother gave birth to him on 7 Dec. 1941."

Here are the four legitimate uses I outlined earlier:

One, to soften an uncomfortable subject, such as "The test animals were euthanized" instead of "We killed the test animals."

Second, to hide credit or blame. "Were euthanized," in the above sentence eliminates the need to say who did the nasty deed.

Third, to inject variation into writing.

Fourth, as a way to keep sentences gender neutral without using "he or she," as in: "If an author who writes in passive voice were to convert most of those sentences to active voice, the total length of the paper could be reduced at least 5%, and perhaps as much as 20% or more . . . ."

Other than those . . .

I have been pleased to see more and more authors experimenting with active voice recently, and fewer and fewer reviewers and editors discouraging its use.

These little drops on the granite facade of our journals will not cause them to crumble, but rather to become polished and much more attractive not only for all who use the journals, but perhaps also for others who have been afraid until now to use them.

* * *

An anonymous person sent me a clipping from a British scientific journal praising passive voice. I wrote a column based on that article and related search on my own, but certain people at ASA–CSSA–SSSA Headquarters did not allow me to publish it in its newsletter, so I published it in my own newsletter, in April 2000:

Scientists make case for passive voice

Although many scientists agreed with my February 1997 column, which did not find much use for passive voice in scientific literature, some said their colleagues did not agree.

Bob Loomis, of the Univ. of California-Davis, wrote: "My Australian and European editors take *I* and *we* as examples of American arrogance. We try to hold it to a minimum."

Keep his phrase, *American arrogance*, in mind as you read from a letter written by Donal P.L. Murphy, a native English speaker in a German research establishment:

"In Germany, scientific writing in the active form (in German) is looked down upon. Some seem to associate it with the style of a farmers' weekly newspaper. Many colleagues ask me to advise them on scientific text (in English) and I have encouraged the use of the active form in some cases. They are often surprised when I explain that traditionally this form is used in the most respected journals such as *Nature*. I believe active voice is useful where one wants to:

- Remind the reader that the scientific work was done by people and not machines.
- Emphasize teamwork. I find it good where groups of scientists use the active form *we*. For me, this underlines the collective responsibility and teamwork that went into the reported work.

"I think also the active form reinforces the purpose of a scientific paper, which is to report observations made by one or more scientists.

"I encourage a mixture of both forms. The active form in an introduction, for example, followed by the passive form in the methods section. I think also the active form can be effectively used in a discussion."

The case for passive

So, depending on the critic you listen to, the use of active voice is either American arrogance or writing for a less-educated audience.

The strongest case I've seen for passive voice (and it isn't very strong) was given by Simon R. Leather, Dep. of Biology at the Imperial College in Berkshire, England. In "The case for passive voice," (*Nature*. 1996. 381:467) Dr. Leather says extremely well-written passive voice is superior to poorly written active voice. No one would disagree with that.

Some of the advantages of passive voice, in Dr. Leather's words:

- Using the passive voice in scientific writing allows the researcher to stand at a distance from his or her work. By standing at a distance an unbiased viewpoint is much more likely to be reached. An unbiased viewpoint encourages a world view and an open mind, surely prerequisite for good science.

- The use of the passive voice encourages disciplined writing, cases must agree, tenses must be used correctly. It is therefore more demanding, but the precision and professionalism displayed is worth the effort.
- The use of passive voice encourages precision and probity, and when used correctly can generate as much passion and stimulation as the skilled user of active voice.

Active voice, in Dr. Leather's opinion, does the opposite:

- Using the active voice is an easy option. There is no need to discipline one's thoughts. An author can just pour out his or her thoughts. This leads to careless presentation, particularly in methods and materials sections.
- Most authors using the active voice show no consistency of use. Papers alternate between passive voice statements and active voice statements, sometimes in the same paragraph with no logic for the change of voice. This results in a paper full of inconsistencies and a general mixture of styles.
- Using active voice engenders possessiveness in the results and/or work. By engendering possessiveness an author risks adopting a biased and partisan stance. It is tempting when writing from a partisan viewpoint to descend to spite and denigration of other work. This too often manifests itself in bias in anonymous peer review of manuscripts and grants. There is also the possibility that use of the active voice and the resulting adoption of results and hypotheses as the author's own personal property leads to an unwillingness to see those results contradicted or refuted. This may, in the worst case scenario, lead to the fabrication of results, something seen much more today than 20 years ago when the passive voice was *de rigeur*.

My reaction

Passive voice alone does not engender great writing and active voice alone does not engender terrible writing. I found Dr. Leather's notion that active voice engenders scientific fraud so bizarre that I asked him to clarify it. His response: "I wrote my contribution from a somewhat controversial position to stimulate discussion. In reality, I am not totally opposed to the active voice in scientific writing. It is the use of *I* or *we* that I feel can, in a few cases, tend to encourage partisan views which can lead to bias in referees, and ultimately, fraud. As an editor of an International journal (*Ecological Entomology*) I have seen many examples of the former, but to the best of my knowledge, none of the latter."

I also find it difficult to agree with Dr. Leather that it is admirable to be dispassionate about one's beliefs. Surely he is not saying that a person like Barbara McClintock was wrong to stand by her findings for decades until her peers in the scientific community finally realized she was correct.

I disagree with Dr. Leather's statement that it is easier to write in active than in passive voice. It is difficult to write well in either. As to his point that poor writers switch from one voice to another and back again, did it bother you (or did you even notice) that I wrote the first sentence of this paragraph in active voice and the second in passive voice? This was not poor writing, I switched so the second sentence would parallel the last part of the first sentence, which was an indirect quote of Dr. Leather.

At another point in his editorial Dr. Leather asks, "How many of the most memorable prose passages in English literature are written in active voice?"

Well, there was, "A horse! A horse! My kingdom for a horse!" That is far more memorable than its passive equivalent, "In exchange for a horse, my kingdom will be given."

"Romeo, Romeo, wherefore art thou, Romeo?" is certainly preferable to, "The location of Romeo is unknown."

Moving from Shakespeare, there is certainly nothing careless about the memorable line in active voice from the sports writer Grantland Rice: "Outlined against a blue-gray October sky the four horsemen rode again."

As for memorable lines in passive voice, Dr. Leather did not repeat any, and the only one a colleague and I could think of was, "A mighty fortress is our God."

Awkward passives

As to passive voice being more logical, consider: "The test was unable to show . . . ." Of course the test was unable to show anything, it is an inanimate object. A more logical phrase would be, "We could not learn from the test . . . ."

A (mercifully) few scientists seem to love to use the verb "was" in their sentences: "The test was conducted but was unable to detect what was the reason for the problem." A colleague, upon reading a sentence like that, remarked, "Nobody did it, it just was."

Or remember the ambiguous passive-voice sentence in my March and

June 1997 columns [final two in Chapter 6 of this book]. The best I could come up with in passive voice was, "Some 49 papers were presented at the recent conference." Even his final version does not tell who presented the papers. The following active-voice sentences eliminate that ambiguity and also demonstrate that such sentences can be written without using *I* or *we*: "Members presented 49 papers at the recent conference." "Scientists presented 49 papers at the recent conference."

Conclusion

I agree with Dr. Leather's comment that an excellently written sentence in passive voice is superior to a poorly written sentence in active voice. He agreed with me in private that an excellently written sentence in active voice is superior to a poorly written sentence in passive voice.

Active voice and passive voice each have their place, as I explained in my June 1995 and February 1997 columns [this chapter]. But most people familiar with the process of scientific writing say the place for passive voice should be far less prominent than it has been in recent years.

* * *

I thought that would be the last word on the subject, but then came across a piece in a newsletter for editors titled *Editorial Eye*, which presented some logical thoughts on one aspect of this picture. The following excerpts from that article seem to agree with me that passive voice is not a good idea, but express reservations about the use of the pronouns *I* and *we* that often (but not always) are used in this voice.

Following are two excerpts from "First person in technical writing" by Jim McDermott, a Ph.D. candidate and writing instructor at the University of Virginia; published in the August, 2000, *Editorial Eye*:

Advantages of I/we

Most of those who give advice about effective technical writing recommend the first person, which means referring to the writer(s) as I and we, chiefly because it leads naturally to—although it doesn't necessitate—the use of the active voice. The active voice clarifies meaning, not only by specifying agents but also by yielding simpler syntax and greater concision than the passive voice; use of agentless passive voice only when the agent is unknown, unimportant, or evident from the context.

Another reason to use the first person is to avoid third-person descriptions such as *the author* that catch the writer attempting to hide in plain sight, apparently for no other reason than to make a self-aggrandizing display of modesty. As H.J. Tichy comments in his detailed and exceptionally generous book *Effective Writing for Engineers, Managers, and Scientists*, "Discarding necessary common words like *I* and *we* merely leads to awkward writing marked by excessive use of the passive voice and by reliance on weak indirect constructions [such as *the present writer*]."

Of course, good-writing mavens, rigorous utopians that they are, sometimes perform a general rather than local service. That is, they inspire us to dream about how the English language could be used, then leave us to judge whether it will be possible to sustain that vision within the editorial communities we consider home. Although well-known peer-reviewed journals such as *Science* have made a point of encouraging contributors to use the first person, it's important to remember that some readers hold expectations about this usage that differ from those held by the authors of effective writing manuals. These expectations will determine whether I or we will help the message get through or ensure its failure.

Disadvantages of I/we

The resilience of the agentless passive, despite widespread opposition, indicates that there are disadvantages to using the first person, however. For instance, the first person could interfere with the convey-ance by inserting a reference to the writer at the precise moment when the reader is trying to concentrate on what the writer has to say.

On a related point, constant self-reference could suggest that the writer suffers from elephantitis of the ego, souring the reader's impression of him as well as diminishing the credibility of the information he wishes to communicate. Modesty was so important a rhetorical virtue to E.B. White that he put the following piece of advice first in his "An approach to style," addendum to *The Elements of Style*. "Place yourself in the background."

Furthermore, as Jefferson D. Bates points out in *Writing with Precision*, a good reason to Avoid immodest diction is that few of us work without the support of other people and of institutions. The first person also creates a warm, human tone that could be viewed as too familiar and thus rude, like the tone of direct-mail letters that boldly include one's first name.

The pros and cons of using the first person remind us that technical documents do more than convey information or attempt to persuade; they also encode the weighty, and at times burdensome, professional practices of established disciplines. These practices constrain writers who remember to keep the intended reader firmly in mind.

That is to say, readers who are used to the agentless passive voice, or who have learned to expect the active voice and the third person, might view what Bates calls "the perpendicular pronoun" as a breach of professional decorum, one that interferes with the work of grasping information and creates resentment toward the writer. Angry readers don't retain information.

So first things first: Is it how well you know your reader or how well known your reader is? As always, only the reader knows for sure, but we have to make educated guesses in the meantime.

On thing is for sure: No writing technique, regardless of how well intended, can be uniformly and thoughtlessly applied across all publications.

* * *

If there were a simple way to wrap up this discussion I would not have needed to present the discussion in the first place. Clearly, as stated early in this chapter, too much passive voice is bad, but there is an important and sometimes essential place for it.

As some contributors to this chapter and many persons I have talked to and dealt with through the years have made clear, active voice to some extent—and the use of the pronouns *I* and *we* to perhaps a greater extent—are not always cherished in scientific writing.

Perhaps, to quote McDermott in the just-present excerpts, I am one of those language mavens who inspire you "to dream about how the English language could be used," and then leave you "to judge whether it will be possible to sustain that vision within the editorial communities" you consider as home.

Come to think of it, that is not a bad position to be in. I don't know who said it originally, but Robert Kennedy used to often repeat the following quote, "Most people see things as they are and ask, Why? I dream of things that never have been and ask, Why not?"

In fact, you could even go so far as to say that might be the theme of this entire book, not just this chapter.

Chapter 14

Plagiarism

I did something quite unusual during the late summer and early fall of 2000. I devoted virtually the entire August issue of my newsletter to a single topic. Apparently readers liked it, though, for they sent letters with enough additional details or questions asking for additional details from me, that I needed to devote part of two more issues to the topic. What topic is important enough to require all that space? Plagiarism.

I must begin the treatment of the subject of plagiarism in this chapter with the standard disclaimer that I am not an attorney and my opinions presented in this chapter concerning legal rulings on plagiarism and copyright should not be construed as legal advice. Further, anyone who is facing a charge of plagiarism or who wishes to bring charges of plagiarism against another should contact an attorney who specializes in that area of legal practice.

While plagiarism has always been a problem for writers, computers have compounded the problem. Before computers, someone who wanted to steal your work needed to rewrite or rekeyboard or retype every word. Now all they need to do is highlight, copy, and paste.

Is it a growing problem? Seemingly so, as the rest of this chapter will show.

Negotiating the literary quicksand also known as plagiarism

Like other forms of dishonesty, plagiarism is easy to distinguish when it is blatant. Clearly, if you take a substantial number of words or the bulk of an idea that had been written by someone else and publish them in your own paper without attribution to the original author, you have stolen that person's work, which is plagiarism by anyone's definition. (Unlike copyright, plagiarism does not apply only to the arrangement of words. See page 288 for other differences and similarities.)

But, also like other forms of dishonesty, plagiarism is not nearly so easy to distinguish at its fuzzy edges. As I'll explain below, fully attributed material can sometimes fall under the definition of plagiarism, while the exact quoting of a string of words without attribution may not.

Most dictionaries define plagiarism as some kind of literary theft or stealing. Clearly, then, for scientists, engineers, or anyone whose professional lives and careers are heavily defined by their publishing, plagiarism is a dirty word with both legal and ethical considerations.

Definition

Dartmouth College's Committee on Sources in 1988 outlined three types of plagiarism:

- Direct quotation or word-for-word transcription
- mosaic or mixing paraphrase and unacknowledged quotation
- paraphrase and/or use of ideas

Of course, to actually be plagiarism, any of the above would have to be done without credit to the original source of the information. I would submit that the above listing is a poor definition of plagiarism because it leaves out the two most important elements.

The test of plagiarism involves at least two things. The first is that the material in question is a usually *unattributed* quotation or paraphrase used with or without permission of the original author of a *substantial* portion of that original author's work. Second, the original author must be able to present evidence that the accused author had an opportunity to read the work from which the material had allegedly been lifted.

Of these two items, the second is the easier to explain, so I will deal with it first. Let's take an extreme example, and say that something that you have written and published is nearly identical to something a person in Timbuktu wrote 5 years earlier and sent in a letter to a friend in the same city. If the author who wrote that letter and the friend who received it were the only two people on earth who ever saw the material contained in the letter, that author has no case against you because there is no way you could have stolen the material.

Such clean examples rarely happen, but there have been cases dismissed because the author accused of plagiarizing another's material was able to demonstrate to the satisfaction of the court that he had no access to or knowledge of the material in question.

On the other hand, cases have been won by authors who were able to discover that the accused author had served as a scientific reviewer of the paper from which information had been stolen.

Attributed material

Perhaps I should dispose of another fairly easy matter up front as well; that of how fully attributed material can be considered plagiarism. If an entire paper says nothing more than "Johnson says this, Johnson says this, Johnson says this," that is not the second author's paper, it is Johnson's paper, and is therefore not a substantial addition to scientific literature on the subject.

Few reputable journals would publish such a paper. But many journals face from time to time papers on the fuzzy edges. Its reviewers need to decide how much quotation of another author's work is too much and if the level of input from the second paper is or is not a substantial addition to the literature already published on the subject.

Defining "substantial"

If I were to write a poem that began, "Oh say, can you see by the dawn's early light what so proudly we hailed at the twilight's last gleaming," that is obviously a direct quote of substantial length from Francis Scott Key, and if I did not acknowledge that fact, I would have committed plagiarism.

If, on the other hand, I were to write, "At first light we could once again see the flag, still proudly flying over the fort, just as it was late yesterday before the shelling began," that would be an account of the same incident in my own words, and would not be plagiarism. (This is one of the reasons why *Time*, *Newsweek*, *U.S. News*, the television networks, the radio networks, and hundreds of newspapers and Web sites can all publish accounts of the same large news story with hardly ever a hint of either plagiarism or copyright infringement.)

Someone writing of that event in words closer to those of Key's, particularly if it could be proven that he/she had read or heard Key's poem before writing those words (and who in the USA and most of the rest of the world is *not* familiar with that poem?), is on shakier ground.

Let's say, though, that someone were writing an article on the restoration of the original flag that inspired Key's poem. It would be quite appropriate to begin that article with the words, "Oh say, can you see how tattered the old banner has become?" That author obviously used the first five words from Key's poem in the exact same sequence. But that would be considered a literary allusion, not plagiarism.

Just as scientists sometimes joke about the "minimum publishable unit," others have discussed the "minimum stealable unit." Unless you invent a word, every word you use in your writing has been used by at least one other author, so reuse of individual words is not plagiarism.

Believe it or not, though, one author tried to protect a single word he had coined. That author was Vladimir Nobokov, who once took someone to court for using the word "nymphette," which he created and used to describe the main character in his novel *Lolita*. (Most likely that particular lawsuit had much more to do with gaining publicity for Nobokov's book than it did an actual charge of plagiarism.) At any rate, Nabokov lost because the court said there is no way to copyright a single word and both copyright infringement and plagiarism must involve the copying of a substantial amount, which is definitely more than a single one word.

But what constitutes "a substantial amount" of the original material? Years ago, *substantial* referred only to length. It took on a broader meaning during the trial of a news magazine that was sued for reprinting the section from President Gerald Ford's memoirs dealing with the pardoning of President Richard Nixon. The news magazine argued that it had lifted such a small percentage of the entire book that it should not be held liable. President Ford's attorneys argued (and won) that even though the stolen portion was a physically small part of the entire book, it was the most important part and the reason why most people would buy the book.

What this means to scientists

The implications of that final point seems, to me, to have a profound significance to scientists because scientific discoveries can sometimes involve quite a small aspect of a large picture. Consider, for example, the impact on plant breeding and many related and unrelated fields of the discovery of hybridization and especially of hybrid vigor.

Or consider the impact the explanation of plate tectonics had on most aspects of not only the physical sciences, but also fields such as archeology, paleontology, oceanography, and even weather and climate studies.

Regardless of the scientific field, every now and then one small discovery will constitute a breakthrough that has an enormous impact on a host of things in that field, and sometimes in a number of related and even unrelated fields. Current genetics studies certainly have that potential. Stealing one small key discovery could constitute a theft of major proportions.

So what does a scientist need to remember about plagiarism? In the words of one author: "Plagiarism can also result from simple ignorance in how to reference other works or in sloppy, careless documentation. Careful documentation and verification of your references as well as clear knowledge of what has been done in your area of study are essential. Learn the accepted conventions for documenting the works of others and for obtaining copyright permission. When you question your right to use the words or ideas of theirs, find answers to those questions before you proceed. Ignorance is no excuse." [Excerpted from M. Davis (1997) *Scientific papers and presentations*, Academic Press: New York City.]

Or, in my words, if a substantial amount of material from one paper appears in another paper without citing or acknowledging the original author, and if it can be proven that the second author had likely seen the first paper because he either subscribed to the journal in question or had been one of the reviewers of the first paper, the first paper's author(s) has (have) a good chance of convincing a judge and jury that the second author had plagiarized the material (stolen the intellectual property).

Questions of credit

Another area where plagiarism is often charged involves disputes among two or more scientists who collaborated for a number of years, then went their separate ways. In their subsequent work they made independent use of the concepts, methods, descriptive language, and/or products they had developed together to come up with new findings.

The precise ownership of the jointly developed intellectual property is often unclear. Even though cases of this sort are generally not considered plagiarism, they can, still involve heated discussions, claims, and counterclaims.

The ownership of the intellectual property in many such situations is seldom clear, but the collaborative history among the scientists often is taken by courts as the presumption of implied consent by both former collaborators to use the products of that collaboration.

Few things are truly new

Surely no person would never cite a thought as "developed from a personal conversation with Jones, who said she was inspired in that belief by Johnson who had read a paper by Snodgrass about a finding of Haverhill in a paper reviewing the work of Abercrombie, who said he was inspired by a random thought from his mother's gardener's grandson."

We also know that as ideas get passed on in this way they are embellished upon by new positive and negative findings.

Still more vague is what to call it when you pick up from your subconscious something you read long enough ago that it has become a part of your own thought process. We all know there are few truly new ideas. Everything else is a new way of looking at, thinking about, or working with something that has been around for a long time.

Indeed, some of the greatest findings of all time have come about by creative use of another's discovery. Eli Whitney invented a way to manufacture muskets on an assembly line and someone else (probably Gottleib Daimler, but who can say for certain?) invented the automobile. Henry Ford, though, figured out a way to manufacture automobiles on an assembly line, and changed the course of history.

Puppets have been around for thousands of years, but Jim Henson created a whole new way to use them by combining the ancient art of puppetry with the new capabilities of television and movie cameras.

What's a scientist to do?

While nebulous situations can occur and everyone uses ideas that are not entirely original, scientists need to give credit where credit is due for words and ideas they publish in their papers. Three points to remember:

- If you directly quote someone else's words, let the readers know what you did and tell those readers who the person was that originally expressed those words.
- If you directly quote (with attribution) someone else's words, then paraphrase additional words from that same or another author without quotation marks, make it clear in your paper that those are paraphrases of another author(s) and not your own take on the quotation just given.

- And always remember the word *substantial*, and that this word can apply to things other than just length.

Whether what you have copied or paraphrased from another constitutes plagiarism depends not only on whether the paraphrase is attributed to its rightful creator, but also whether your paper contains so much of it that the bulk of your paper is a reiteration of the original material instead of a presentation of new material by you.

Most scientists who regularly publish are familiar with the literature in their field and generally know if the thing they are writing in their own paper is only barely from someone else or if it had been lifted substantially intact from something or someone else.

Charges of plagiarism are far from new

Plagiarism (or at least the suspicion of it) has been around for a long time. For about as long as people have been writing, other people have accused them of this crime.

Some claim that Milton lifted many passages of *Paradise Lost* from other poets. Others contend (but have never convinced me or most other people) that Shakespeare wrote little, if any, of the material attributed to him.

More recently:

- Alex Haley paid a large settlement to the author of a book titled *The African* who contended that Haley had lifted large segments of that book and published it as his own in *Roots*. (At the trial, Haley said he didn't personally do it, and blamed one of the researchers he had hired.)
- A large lawsuit was filed against Steven Spielberg by an author who claimed his movie *Amistad* used much of her book without credit or compensation. That has not been settled when this was written.
- The creators of the TV series *Star Trek Deep Space Nine* paid a settlement to the creator of the TV series *Babylon 5*, who claimed that he first pitched his series to them. After they rejected *Babylon 5*, he began to develop it on his own. Before he could get it on the air, though, they brought out *Deep Space 9*, which was strikingly similar to his.

In short, there is a lot of suspicion of plagiarism going on and at least enough real examples of it happening to keep people suspicious.

Academic situations

Virtually any scientist who has been in the field for more than a few years and virtually every professional in the field of scientific publishing for more than a few years can recite at least one example of suspected, if not charged, if not actual plagiarism.

One that has been told so often that it must have at least a grain of truth in it concerns a graduate student who, in the course of his or her studies, achieves a breakthrough in a particular technique or aspect of something. The major professor is eventually able to capitalize on this discovery and, several years after that graduate student has received the degree and moved on, publishes the paper that was possible only because of that breakthrough without mentioning the name of the person who actually made the breakthrough.

Differences between plagiarism and copyright

Copyright concerns temporary legal ownership of intellectual property, plagiarism refers to perpetual ownership. While these two terms overlap to some extent, they are far from synonymous.

Before enactment of the current U.S. copyright law in 1976, authors or publishers could copyright their material for 17 years and renew that copyright for an additional 17 years. Doing this gave these copyright holders full say concerning the use of their material (and allowed them to collect royalties for that use) for up to 34 years. After that, the material moved into what is called *public domain*, which means that anyone can use the material without paying royalties to the former copyright holder (although the original author must still be credited).

Hugh Hefner took advantage of this fact when he was beginning *Playboy* with lots of dreams and little money. Virgil Partsch was one of the most famous cartoonists of that day. Hefner was aware that many of Partsch's cartoons were more than 34 years old, so used those and work from other famous cartoonists and authors that had gone into public domain. It allowed Hefner to publish quality material from famous contributors in his fledgling magazine (enhancing its quality image) without the need to pay royalties for that work.

The current copyright law extended the time limit on copyright ownership to the life of the author plus 50 years. Further, under the law authors are assumed to possess a copyright on everything they produce unless they specifically sign it over to someone else (or are U.S. Government employees or employed under a "work for hire" or other exempt situation).

What this means, for instance, is that even if some month I forget to put a copyright symbol on my newsletter, it is still legally copyrighted by me and cannot be reproduced without my permission (or the permission of my estate) until 51 years after I'm dead.

As under the old law, after that time the material moves into public domain and can be used by anyone without paying royalty to the former copyright holder, *so long as the author still gets the credit*. Therefore, I could assemble and sell a book of sonnets by Shakespeare without having to pay a royalty to anyone. But, if I were to say that I, and not Shakespeare was the author of those sonnets, that would be plagiarism, for which there is no time limit.

Sources of information on plagiarism

I thank Clay Robinson of Texas A&M University at Canyon for suggesting this topic to me. When I received his suggestion in an e-mail I went to the Web, entered "plagiarism" in the Lycos search engine, and learned in a couple seconds that it had 32,454 listings on the subject. A couple more mouse clicks and I found that "Amazon.com" lists 63 books for sale on that subject. I even found several companies that offer software that is supposed to detect plagiarism in a piece of writing.

Some listings, particularly from universities across the nation and around the world, carried substantial amounts of material; apparently a sort of a short course for students on the subject. Several gave examples of a quotation from an original paper followed by two or three paraphrases, some of which it considered plagiarism and others it did not.

What I could gather in my research is that the widespread availability of intellectual property on the Web (and the ease of downloading that material) has increased the amount of stolen material being found in student papers across the nation.

If this is a concern where you work and you want to learn more on this subject, there is plenty more to be found. In addition to the material already presented in this month's newsletter, here are a couple more quotations you may find as interesting as I did:

"Plagiarism is easy to commit. It takes neither talent nor ingenuity to copy another person's work and try to take credit for it. However, 'getting away' with plagiarism may be much more complicated. This is because people who are likely to read your work are also very likely to have read what was plagiarized.

"When that happens, the best you can hope for is that your integrity will be diminished in the view of your reader. And the worst that can happen may be very bad indeed. . . . It can also lead to sanctions by funding agencies, up to and including, in the U.S., time in prison if it can be proven that the scientific misconduct caused a waste of federal funds. . . More importantly, you will have broken faith with the rest of science."

— *Univ. of Texas-Houston Graduate School of Biomed. Sci.*

"Although there is widespread agreement in the scientific community on including plagiarism as a major element of the PHS definition of scientific misconduct, there is some uncertainty about how the definition of plagiarism itself is applied in ORI cases.

"As a general working definition, ORI considers plagiarism to include both the theft or misappropriation of intellectual property and the substantial unattributed textual copying of another's work. It does not include authorship or credit disputes.

"The theft or misappropriation of intellectual property includes the unauthorized use of ideas or unique methods obtained by a privileged communication, such as a grant or manuscript review. Substantial unattributed textual copying of another's work means the unattributed verbatim or nearly verbatim copying of sentences and paragraphs which materially mislead the ordinary reader regarding the contributions of the author. ORI generally does not pursue the limited use of identical or nearly identical phrases which describe a commonly used methodology or previous research because ORI does not consider such use as substantially misleading to the reader or of great significance."

— *Office of Research Integrity, U.S. Public Health Service*

Is it ever possible to plagiarize yourself?

Sir:

I have just finished reading your article on plagiarism and have this question. If I lift a paragraph from a previous paper that I authored, let's say a method description, and use it verbatim in another paper (without citing my previous paper) is that plagiarism, copyright infringement, or a time saving device? I realize I could just cite the method from the previous paper, but there are times you want to give the reader information without forcing them to read your previous work to get it. So I guess my basic question is can you plagiarize your own writing?

— *J. Michael Kelly*
Agronomy Department
Iowa State University
Ames, IA

My reply:

First of all, a scientist may not publish the same data twice and should not even submit it simultaneously to different journals. But you are not talking about doing that. I agree with you that the presenting of a bit of procedural material that may have been published elsewhere could benefit your readers.

Using a reference to an imaginary paper, here are three ways you might present information such as you refer to in your letter.

1. The tea was made as described in Luellen (1972).
2. The tea was made by boiling the water for 10 s, immediately pouring it onto 20 mL of tea leaves in a cup, allowing the mixture to brew for 5 min, then removing the leaves.
3. The tea was made by boiling the water for 10 s, immediately pouring it onto 20 mL of tea leaves in a cup, allowing the mixture to brew for 5 min, then removing the leaves (Luellen, 1972).

The first of these three is what you wanted to avoid, the second is your suggested solution, the third is mine, using the same format.

Using the same idea as the third, just presented, here is perhaps an even better way to present the sort of information that is more likely to appear in a paper you would write: "As reported previously (your name and year), but repeated here for the convenience of the reader: . . . ."

I would limit such quotations to methodology and other sorts of things that are short enough to cause a major nuisance to a reader who was forced to go elsewhere to look it up. How short is that? I don't know; a couple paragraphs, maybe.

Before publishing it, I ran the above reply past a number of scientists who are either presidents of scientific societies or editors of scientific journals. All who responded to my e-mailed query (a majority of those polled) agreed with my reply.

Now for your other question (is it plagiarism or copyright infringement), in a word, no. Even if you have signed away your copyright of the piece, the law allows brief quotations from copyrighted material for "fair comment, and criticism" and for scholarship. As you know, most journals do not ask authors to sign away their ownership of a research paper, but instead to give the journal "Permission to Print and Reprint" that work.

Once you step beyond scientific writing and publishing you can see authors "plagiarizing themselves" all the time.

The easiest place to spot it is in music. For instance, the entire second melody in Gershwin's *Rhapsody in Blue* (the one used in the United Airline commercials) was lifted intact from an earlier piece Gershwin had written. And that is but one of many examples. Anyone familiar with more than a few works of Beethoven, Mozart, Bernstein, or virtually any composer will recognize recurring themes and melodies.

As for writing, Shakespeare often recycled quotes, as did Emerson, and many authors.

To repeat, scientific writing is a special case. Scientific authors need to think not only of plagiarism, but also of multiple publishing. Depending on the way you treat the information you want to present, it might or might not be considered multiple publishing.

Repeating substantial amounts of data is not allowed. But repeating conclusions drawn from that data, methodologies devised from earlier research, and perhaps even small amounts of data to save readers from having to go back and dig it out—all the above clearly stating what you are doing and citing the original source—would not be considered multiple publication in any case I can think of.

More on plagiarizing yourself

Sir:

I do not believe it appropriate to use parts of one paper in another paper unless citing it as such as a quote. I know of a case where a person reproduced the exact table from one paper in another. He was charged by colleagues with fraud and was investigated for fraud by his university because of this mistake in judgment. The finding was that it was not fraud, but it caused a local investigation and could have become a matter requiring notification of the federal agencies if fraud was suspected by the local panel.

[name withheld at author's request]

Chapter 15

Reviewing and Reviewers

Every published author and every author who hopes eventually to publish something either has or will soon deal with one or more reviewers. Moreover, every author who has published more than a few things has been called upon to review the work of another author.

From the second, or perhaps third of my original columns I tried to keep the above facts in mind. A number of columns were not as much on the subject of writing as they were on how to review the writing of others. And, of course, many more columns were written on the subject of how to handle comments and suggestions from reviewers.

This chapter is divided roughly in half. The first three columns deal primarily with reviewing papers written by others. Is this placing the cart before the horse? I don't think so. My reasoning for dealing with reviewing, before being reviewed, is to give authors who may be unfamiliar with reviewing a bit of a walk in the shoes of reviewers. My hope is that it will also help answer some questions they may have had as to why reviewers treated their papers the way they did.

With that background out of the way, the final four papers in this chapter will deal, in one way or another, with how an author should deal with questions and suggestions from reviewers.

The first column I wrote on how to review a paper was published in March 1993, and updated for this book:

When you're asked to review a paper

Review of a paper is part of the writing-reviewing-editing-publishing process. Whether you write a paper yourself and have it reviewed by your peers or are called upon to review someone else's paper, you can benefit from an understanding of the review process and the difference between a good and a bad review.

Purpose of a review

Maeve O'Connor, a scientific editor from Great Britain, once defined the purpose of the review-editing processes as: "Protect readers from authors and authors from themselves." Her definition is similar to a humorous one I often use: "We are here to make you look good." Ideally, a good review will do what O'Connor and I have said. If you ever have experienced bad reviews, you know what they can do.

Ideally, reviewers have one thought in mind: "How can I help this become the best possible paper on this subject?" Unfortunately, some reviewers seem to be thinking: "Can I find enough flaws in this paper to prevent it from ever being published?"

The next time you review a paper, instead of taking revenge on everyone who has given you bad reviews, think of the Golden Rule and review a paper the way you would like to have your next paper reviewed.

Avoid bias

You probably should not review a paper if there is the possibility of (or even the appearance of) a conflict of interest. The following seven possible conflicts were compiled by [then] CSSA Editor-In-Chief P.S. Baenziger. These should be considered the minimum:

- Have you had significant and acrimonious disagreements with the authors?
- Are you and the authors co-investigators on a current research project?
- Have you and the authors jointly published an article in the past 5 years?
- Are you a close friend of any of the authors?
- Are you working in the same area of research as the authors so that you might be considered to be a competitor or to gain an advantage by reviewing this manuscript?
- Are you at the same location as the authors?

- Did you review and approve the manuscript as a departmental or agency peer review prior to its submission to the journal?

Confidentiality

A paper in review is the property of the authors or their employer. You violate international copyright laws if you retain a copy of the manuscript or, in any way, make use of the information contained within.

Authors occasionally report to us that their manuscripts are returned with staples removed and replaced (possibly indicating that they have been photocopied) or that they overheard rumors at a scientific meeting that their paper had been passed around to all the researchers in a department. Some reviewers have even refused to return a manuscript copy because they wanted to contact the authors to collaborate with them in their work. If any of these rumored happenings actually happened, they would be violations of both copyright laws and also professional ethics.

The peer review process is anonymous. By that we mean that ASA, CSSA, and SSSA editorial board members and the professional editing staff at Madison will do everything in their power to prevent authors from learning the names of the people who reviewed their paper. Not only that, we also do not let one reviewer know the names of the other reviewers on that paper; although we sometimes will make copies of the reviews of a paper available to all its reviewers as a teaching tool. When that happens, reviewers get only the copies of the reviews, never each other's names.

Even if you ask us to tell the author you were one of the reviewers, we will not. If, after a paper is published you would like to reveal to its authors that you were one of the reviewers, that is up to you.

Some of our journals are experimenting with a double-blind review. With this, the authors names are removed from the manuscript so that the authors are (supposedly) unknown to the reviewer as well as vice-versa. (Several reviewers have said they could pretty well guess the author from the location and subject matter.) It is too early to say how successful double-blind reviews will be, but early indications seem to be more positive than negative. [Later reviews were essentially all positive, and the process was permanently adopted by two of the Societies' journals.]

A good review

The eight qualities of an excellent review, according to the SSSA *Reviewer's Guide*, are objectivity, accuracy, relevance, thoughtfulness, explicitness, helpfulness, courtesy, and promptness:

Objectivity: Strive to make your evaluation strictly on scientific grounds; free of personal comments, professional jealousies, or bias. Subjective statements ("I think this paper is worthless") are inappropriate.

Accuracy: Authors are responsible for the accuracy of the paper; reviewers are not expected to attempt to verity the results. If, however, something does not seem accurate, you should express your uncertainty.

Relevance: Confine your comments to an evaluation of the subject *matter in the manuscript.* Do not make comments about material you think should be included unless that material is essential for the establishment of the points that are included.

Thoroughness: Reviewers are frequently criticized for submitting comments based on a superficial reading of the manuscript. Study the manuscript thoroughly before you make comments. Support your comments: for example, if you say the manuscript contains nothing new, provide literature citations of the similar published work.

Explicitness: Be precise about your comments. Instead of saying the paper is poorly organized, tell what you don't like about the organization and suggest how it can be improved.

Helpfulness: Again, don't just say what is wrong with the paper, suggest how it can be improved.

Courtesy: Write the kind of comments you would like to see in the review of your next paper. Using anonymity to make nasty, sarcastic, personal comments is never appropriate. Editorial board members are instructed to delete them or to not forward reviews that contain inappropriate comments.

Promptness: The seemingly longest time in the world is that between submitting a paper and learning how, or if, it survived the review process. You don't like to be kept waiting on your papers, so don't subject authors of the papers you review to the same torture. If you cannot review a paper and get a response back within a couple of weeks, send it back to the editor immediately so it can be sent to a reviewer who has sufficient time.

Practical suggestions

The two overall questions to ask when reviewing a scientific paper are, "Is the research worthwhile?" and "Does this paper present enough information to allow a competent scientist to repeat the experiment?"

For a review paper, ask, "Is this an accurate, unbiased, and thorough review of the subject matter?"

The question for a "Notes" paper is, "Does this accurately explain the research technique, apparatus, or unique phenomenon described?"

Finally, here is a list of things to guide you in a review:

- Does the title accurately describe the contents without being too long?
- Does the abstract tell the rationale and objectives for the study, the materials and methods used, and the results and conclusions as specifically as possible in the limited space available? The abstract should be complete; statements such as "will be discussed" and "are presented" are not appropriate in an abstract.
- In the literature review, does the author give credit to relevant contributions of others and place them in proper perspective?
- Are the stated objectives adequate and appropriate in view of the subject matter?
- Are the methods appropriate for the purpose of the research? Have suitable measurements been performed? Have proper control measurements been used? Do the statistics clearly establish the results (regardless of whether they are the statistics you think should have been used)?
- Is the material presented in a reasonably straightforward way that can be readily understood by a competent reader. (Keep in mind that readers will include graduate students and scientists whose first language is something other than English.)
- Is the manuscript well organized; does it develop the subject logically and effectively; are S.I. units used throughout?
- Does the manuscript unnecessarily repeat published work? Can the manuscript be shortened without loss of content by condensing two or more tables into one? Are the tables and figures required? Is there unnecessary duplication within the text or between text, figures, and tables?
- Do the calculations appear to be accurate? You might select a few at random and try to verify them.
- Would material now in a graph be more understandable in a table or vice-versa?
- Are all the tables and figures referred to in the text?
- Do the titles and captions of tables and figures accurately reflect their contents? Remember, most scientists will read the tables and figures before they read the text of the paper.
- Are the conclusions accurate and supported by the data? If there is conjecture, is it clearly distinguished from fact and appropriate, based on the facts presented?

- Are there obvious discrepancies between text and references; names misspelled in text or reference list, references not cited, or citations not referenced?
- Is ASA–CSSA–SSSA editorial style used throughout?

Please note that style was the final, not the first item in this list. If the paper is worthwhile, complete, and accurate, Headquarters can take care of style items you may have missed. But Headquarters cannot judge whether the science is worthwhile, complete, and accurate.

When you are reviewed

Keep in mind that you are the author of your paper. The reviewers' suggestions are just that. All reviewer suggestions must be responded to, but they do not have to be heeded. A legitimate response can be, "I did not incorporate and second comment of Reviewer 3 because . . . ."

The editor, of course, has the right to determine whether your refusal was appropriate. If you present your case well it is you, not the reviewer, who will prevail. Keep that in mind too, the next time you see a published paper you reviewed that does not incorporate all your suggestions.

* * *

The next two were published as a pair in March and April of 1996:

Be careful what you change when you review or edit a paper

Editorial board members and ad hoc reviewers often ask me what they should look for in a paper; which sort of changes and comments are appropriate and which go too far.

Such a question is not easy to answer, for it depends on the skill level of the author, the nature of the paper, and the place where it will be published. Even if both were written for the same journal, the first scientific paper from a graduate student will require not just a higher level, but also a different kind of scrutiny than a guest editorial by a retiring officer. The basis of the first paper is facts, while the basis of the second is opinions. Any paper by an author who does not normally write in English will need a higher level of scrutiny than either of the above, regardless of the nature of the paper or where it will be published.

There are, of course, a few rules that apply to all scientific papers. Was the hypothesis correctly established, accurately tested, and adequately defended? Was the presentation complete enough so the work could be duplicated by a competent scientist?

Obvious errors in statistical treatment, math, science, or presentation must be questioned. If the author writes, "Three things must be remembered," then lists only two, you need to ask the author whether to change the "three" to "two" or what that third thing would be. If an author writes a more complicated version of "two plus two equals five" you need to find out if the typist left something out of the formula or if it was simply a calculation error.

Next month's column will expand on the previous paragraphs and discuss specific scientific and technical things to look for in other people's (and your own) writing.

Levels of scrutiny

Once the accuracy of the logic and the science are correct, there are other considerations. Should a reviewer or editor correct only obvious style and grammar errors? Should they substitute or delete words and phrases to improve content and flow of material? Should they rewrite sentences and paragraphs? Those three questions define the three levels of editing; (i) style and grammar editing, (ii) substantive editing, and (iii) creative editing (also sometimes called revising or rewriting).

Generally speaking, the only style questions scientific editors and reviewers need to consider are scientific style, such as using molar instead of normal nomenclature in chemistry and use of S.I. measurement units. The author, of course, has the right in any ASA–CSSA–SSSA publication to retain other measurement units parenthetically after the S.I. units.

Beyond scientific accuracy and technical style, it is usually better to leave the rest of the Level (i) and essentially all of the Level (ii) and (iii) changes to your professional editors at Headquarters. Even with seemingly minor changes, there is a fine line between correcting errors and improving the writing in a paper (which are good) and (i) changing the meaning or intent, (ii) damaging the flow, or (iii) substituting your opinions for the author's (none of which are good).

I have wanted for some time to write about these pitfalls, but hesitated because it requires dissecting one person's editing of a second person's writing, which could embarrass both. My dilemma was solved when someone sent me suggested changes to the published column, "How to review statistics in a paper," which appeared in the November 1995 issue [and in Chapter 12 in this book]. I will reveal only that the would-be editor serves on the editorial board of a scientific journal. Though with the best of intentions, the would-be editor made all three types of errors mentioned at the end of the previous paragraph.

The next three sections will discuss and show examples of those error types. Because of the nature of the original work, not all the examples are the same as one would encounter in a scientific paper, but the underlying premises are appropriate to all writing.

Changes in meaning

The would-be editor crossed out the phrase "I hear" in the column's first sentence: "The most common complaint I hear from authors is, 'Our journals are too rigid.'"

The sentence is untrue without that phrase. I have no idea what most common complaint authors might make to their department head, my boss, Society officers, editorial board members, and perhaps their own spouse and families. The only thing I could truthfully write was, "The most common complaint *I hear* from authors . . . ."

The would-be editor also crossed out the first word in the second sentence: "Unfortunately, there is some truth to these complaints . . . ." While the sentence would still be correct without that word, half its meaning is lost. Instead of, "This is true and I don't like it," the revised sentence reads "This is true."

If you believe there is no place for opinion in a scientific paper, I have two answers for you. First and most basic, these columns are not scientific papers. The deeper answer is explored in the article on scientific speculation [published in Chapter 3 of this book].

Changes in character

The third example sentence: "I don't know who first said or wrote it, but I've seen the following posted in many editorial and publishing offices: 'The strongest drive is not love or hate, it is one's need to change another's copy.'" The would-be editor deleted its introductory phrase "I don't know who first said or wrote it, but," and substituted "(anon)" after the word "copy" at the end.

The next revision was similar, but goes further toward substituting the would-be editor's way for my way of presenting information:

[My version] "Scientists follow this dictum as frequently as anyone else, and each has an opinion of how a sentence should be structured—use passive voice, don't use passive voice; always place a comma here, never place a comma here, etc."

[Would-be editor's version] ". . . and each has an opinion of how a sentence should be structured—use passive voice or not, always place a comma here or not, etc."

The revised first example sentence in this section has the same intrinsic meaning as my original, and the revised second sentence is somewhat close to the original in meaning. That said, the first version of each is in my voice, and the second is in the would-be editor's. I'll write more about this below under the subhead, "Whose work is it?"

Changes in opinion

The final example is a sentence the would-be editor tried to simplify by splitting it in two. I wrote: "Authors who were forced to make unnecessary or improper changes to get their paper accepted resent it when a member of your professional editing staff in Madison asks them to make legitimate changes. More than a few authors wonder if their paper will *ever* get published."

The would-be editor rewrote the first sentence to read: "Authors may be forced to make unnecessary or improper changes to get their paper accepted. They resent making legitimate changes requested by the professional editing staff in Madison."

This revision has two shortcomings. First, it is not clear that the improper changes mentioned in the first new sentence were the cause of the resentment mentioned in the second. Second, perhaps the would-be editor believes that *all* authors resent making legitimate changes requested by your professional editing staff. The would-be editor is entitled to that opinion, but has no right to present it so it looks like it is my opinion.

Whose work is it?

What can we learn from the above examples? First, just as there is really no such thing as "minor surgery," there is no such thing as a "minor" editing change, especially once you move beyond the level of style or grammar editing. This is not to suggest that changes should never be made, but that they should be made only with caution.

Maeve O'Connor established a dozen excellent rules in her book, *How to copyedit scientific books and journals* (1986, ISI Press, Philadelphia, PA). The first three are appropriate to this discussion:

1. Leave well enough alone
2. Find good authority for every change
3. Make essential changes only
    a. Is a change really necessary here?
    b. Why is it necessary?
    c. Is my version an improvement or a backward step?

A statement I often tell myself as I edit someone's work captures the essence of these rules: "This isn't the way I would have done it, *but I didn't do it.*" If there is nothing really wrong with it, leave it the way it is.

If, however, you believe a change is needed, by all means suggest it and see if the author agrees. Most changes I have suggested to authors were well received, with a typical response being, "Of course! I should have caught that myself!" More than a few of these authors have gone beyond that to thank me and say they could not have written as well without my help. One went to far as to say, "You certainly changed my sow's ear into a silk purse."

If you have a valid reason to suggest a change, by all means suggest it. If you don't have a valid reason, remember that authors would rather follow their own whims than be forced to follow someone else's.

* * *

What editors and reviewers look for (or should, anyway)

This is a follow up to last month's column, advising ad hoc reviewers and editorial board members what to look for in a paper. Authors also will find this information valuable; knowing what will be checked can help authors make sure their paper is correct before they submit it. I thank the entire professional editing staff in Madison for input to this column.

Where not to help

If you don't know, don't guess. For example, ASA–CSSA–SSSA and many libraries use abbreviations for journal titles that are defined in the Chemical Abstracts Service Source Index (CASSI). If you know the proper abbreviation for a journal title, by all means use it; you will save someone else the time and bother of abbreviating it. The more common abbreviations are frequently published in the reference sections of the papers in each issue of our journals.

If you cite an obscure journal and can't find its abbreviation, retain its full title in your reference list. It takes us only seconds to find a journal title in CASSI and properly abbreviate it. When we begin with an abbreviation we do not recognize, we look it up to see if it is correct. If we can't find it, we guess the full title so we can look it up. If, after several tries, we cannot guess the title we must query the author for the title before we can correctly abbreviate it.

Do the best you can with punctuation, but again, especially for reviewers and journal editors, leave it alone if you are not sure.

In the beginning

Readers typically scan a journal's table of contents for papers that appear to be of interest, and then read the abstracts, tables, and figures of those papers before deciding whether to read the rest. If you want your paper to be read, pay attention to its title, abstract, tables, and figures.

The title should contain no fewer than five and no more than 12 words (not necessarily including prepositions and articles). Don't begin a title with low-impact words. It's hard for a reader to find your paper in the table of contents if its buried among 25 titles beginning with "effects of" or "influences of."

Your abstract will be published without the rest of the paper in secondary journals and other places, so it should be complete within itself and not contain phrases such as "will be discussed." An abstract normally should contain five sections: Rationale, objectives, methods, results, and conclusion. In short, an abstract is a miniature version of the paper.

Abstracts should not exceed 250 words for full papers and 150 words for Notes because many of the secondary journals mentioned above will simply truncate it, even in mid sentence, if too long. Abstracts substantially shorter than the limits are usually too incomplete to stand alone.

The length limit is an average, not a firm number. Obviously if all 250 words are the length of *formaldehyde* that abstract will take up a great deal more space than if all 250 words are the length of *corn*. If your abstract happens to contain 251 or so words, but is not unusually long, there will not likely be a problem with its length. If your word-processing system can give you a character count, 1,500 characters should be your goal, with 1,750 characters your maximum, for a full paper's abstract. The ideal count for a Note's abstract is 900 characters with 1,050 as your maximum.

Tables

Make sure there's a head for the stub (the leftmost column that labels the rows). It's often omitted, but isn't always obvious from the table head.

Make sure there are no units missing from columns in a table. This doesn't happen often, but when it does it slows things down in later stages of review and editing.

Figures

Where a graph has contrasting symbols or lines, try to define them in a legend within the figure itself. If they must be defined in the caption, do it in terms of words, not reproductions of the symbols. The typesetter probably will not be able to duplicate the symbols. Examples of such wording in a caption: solid and open (or hollow) symbols; dashed, dotted, solid, dashed-and-dotted lines; hatched, solid, and open bars.

If you use error bars, be sure to explain them in your caption, for they can have several meanings. If you are an editor and see that a figure lacks a necessary indicator of data variability, such as error bars or an LSD value, have the author correct it before sending the paper to Headquarters. This is not always easy to add in the caption, and many authors are reluctant to make up new figures after a paper has been accepted.

Many figures are now laser-printed. Please be sure to send us the originals, and not copies. Even a good copier can reproduce gray and leave dots and streaks. If these appear within the illustration they can be difficult to impossible to delete, which causes delays while we wait for clean originals from the author.

Never write on the publishable part of a figure or apply a low-stick adhesive note paper to it, even if you think it is only a copy. If you must attach a note, place it outside the image area. Ink, toner from the laser-printed page, or dry-transfer letters or symbols can sometimes come off when these low-stick note papers are lifted, and tapes with stronger adhesives cause even more damage.

Finally, if you are an editor who has marked changes, please let us know whether you have cleared the changes with the author.

Nomenclature

The paper should include the common and chemical names for farm and lab chemicals (reagents, pesticides, etc.). In most journals these names are given parenthetically after the first mention; some journals give them in a separate list. (Contact the journal you plan to submit to if you are not familiar with its procedures.)

The trade name or manufacturer should be indicated only where it is essential information. Some authors cite manufacturer and location for every item used, unnecessarily cluttering the text. Others casually drop trademark names, leaving it to the reader to figure out what the equipment is: "We used a Veebafetzer to obtain the needed information." That's unkind to a cross-disciplinary international readership. Where specific manufacturers or products are named, ask the author if the information is essential to understanding or replicating the work. If so, get the manufacturer's name and location (city and state); otherwise, substitute a generic name or description.

Where the author has given only the scientific name for a plant or animal, please ask for a common name. Include the authority with the name for plants, pests, and pathogens; but not animals.

Most journals have correct listings for scientific names of the most common material they publish, so you don't have to send a paper back just to fill in the scientific name for corn or wheat, for instance, in a crop journal. Please supply or ask the author to supply the scientific name for exotic items, however. Editors will check it against their reference lists, but it helps to know where to start.

Provide the scientific name for the causal agent of plant diseases. Editors will check it against their lists from the American Phytopathological Society, but need a starting point.

Provide the scientific name for insects, nematodes, and similar pests. Again, editors will check it against lists (from the Entomological Society of America) and reference books, but need a starting point.

For soil descriptions, if you have the series name, give the complete description as well. If you don't have the series name, give at least the great group (e.g., Typic Hapludolls). For soils outside the USA, try to get a U.S. great group equivalent to the local description.

Convert normal to molar values for concentrations. Editors have a short list of 1:1 conversions, but otherwise must contact the author. Sometimes the replies are hurried and careless; at least two widely different conversions used for the same substance.

Aside for exceptions allowed by your publisher's style, do not use ppm or ppb; convert to acceptable units such as mg kg^{-1} (1:1 equivalent to ppm) or μg kg^{-1} (1:1 equivalent to ppb in non-British English).

Convert lux to whatever photometric units are appropriate. Many publishers have charged their editors with keeping lux out of their publications.

Check to be sure that N–P–K values are in the elemental, not oxide form. If the author did not write out N_2O or P_2O_5 (or the like), editors cannot always tell if the values represent the oxide form.

Be sure the base is specified for growing degree days.

Watch out for less common units outside of our accepted list, some of which we cannot reliably convert at headquarters.

Statistics

Question statistical inconsistencies in the prose, but change them only with caution and after consulting the author. The level of significance in a recent paper was set at 0.05. At one point the authors said a character "tended" to be such-and-such, and added a P level of 0.10 in parentheses. Does this mean that they accepted the new level of significance, or only that they wanted to let the reader know the basis of their perception of the tendency? To further complicate matters, a scientific editor or reviewer changed *tended* to *was*, making it a definite statement of significance, and did not let us know if the author had been consulted.

Most statisticians say it is sometimes acceptable to report nonsignificant differences to let the reader know what was happening. Such reporting might be relevant in other circumstances, such as or for planning other experiments, or with other significance cutoffs.

Papers may report data using all three of the r-coefficients: r, r^2, and R^2. Occasionally, this is done inconsistently, with one form used where the other was intended. It is hard for editors to check this. More than one author, when queried whether r^2 or R^2 was intended where both forms were used in a similar context, replied that it doesn't matter, that the two are interchangeable, which isn't true. We have also sometimes seen P and p used interchangeably.

Journal style calls for converting percentages to decimals for significance levels (e.g., '$P \leq 0.01$' instead of 'significant at the 1% level'). If you want such a conversion for any other statistic (e.g., 95% confidence interval), please instruct us specifically. We don't routinely convert other statistical percentages.

In tables, if means separation is shown, be sure there is a note specifying what is being compared. It's not always enough to say that means followed by the same letter are not significantly different at the specified level: the reader also must be told within what parameters the comparison is made.

Equations, mathematics, and models

Equations should be in standard mathematical form, not programming language format. All variables in an equation should be defined; they aren't in about one-third of the papers we see with equations. If vectors are involved, please mark vector variables separately from other variables, or at least make up a checklist. Vectors are to be set in bold type and regular single-letter variables in italics. Editors must know which is which so they can give the correct instructions to the typesetter.

Where a computer model uses one form of variable and the mathematical representation would typically use another, decide which will be more helpful to the reader. The mathematical form is more universally understandable and better shows the relations among variables. For example, T_{uc} and T_{lc} for temperature in the upper and lower canopy will relate easily to T_{max} and T_{min} for atmospheric maximum and minimum temperature; UPCANTEMP, LOCANTEMP, MAXTEMP, and MINTEMP are just as precise, but less readable.

Mathematical form is probably better when there are many equations. If the discussion is primarily of interrelationships among the variables or their use in parts of the model, the computer form may be better. If the paper has an appendix or table of variables, both forms can appear there.

Unpublished or personal?

Unpublished data are from the authors' own work; personal communications are unpublished data communicated to the authors by someone else. Both should include the year.

Include initials with the surname for personal communication citations, since this information will not appear in the reference list. Some authors like to add city and state and even full institutional identification. We require only the name and year, but some of our staff editors retain the additional information (except position titles); on the theory that it takes only a little extra space and may help the reader evaluate the source.

In general

If you are an author, try to have your paper complete before you submit it. If you are a scientific editor or reviewer, concentrate on the substance and details of the science and mathematics. Anticipate questions that might be raised further down the line (How should this be typeset?) and try to help the author create a package that can be published with maximum speed and minimum hassle.

* * *

As mentioned in an earlier chapter, I often insert obscure items to entertain trivia buffs. Fans of the old *Mad* comic book may recognize the word Veebafetzer in the previous column as coming from it.

An excellent book to have for all who find themselves in the position of editor is *The Chicago Manual of Style* (University of Chicago Press).

As promised at the beginning of this chapter, now the emphasis shifts from reviewers to authors. Does an author ever dare say "no" to a reviewer suggestion? Absolutely, as I made clear in January 1996:

When to say no to reviewers

All reviewer comments and questions must be responded to, but they need not always be followed. Authors have the right, and sometimes the duty, to respond with a statement like, "I did not incorporate the second comment of Reviewer no. 3 because . . . ."

Some authors have the incorrect notion that the *reviewer* has the final say as to what goes into and does not go into a paper. The *editor* has the final say on the contents of a *journal*, but that editor will be influenced far more by the author than by the reviewer regarding the contents of an *author's paper*. The reason for this is simple; the author's name, not the reviewer's, name appears at the top of the paper. The author is the person who must defend the contents of the paper.

Virtually every scientific journal uses peer reviewers. (The British medical journal *Lancet* is one of the few that does not.) Peer reviewers provide valuable input, but are not some all–seeing and all–knowing monolith whose every whim must be unquestioningly heeded if an author ever hopes to get published.

What should an author to do about reviewer suggestions? First, remember that they are *suggestions*, not requirements.

Suppose a reviewer suggests that you include a discussion of how the findings of your work relate to the work of Dr. X. You have every right to answer along the lines of, "I didn't include a discussion on the work of Dr. X because it is not relevant to the findings of my work. That work involved the physiology of seeds and mine the physiology of mature plants. Dr. Y (appropriate citation for the work) has demonstrated that the two are not related as far as this matter is concerned." If the editor supervising the review of your paper agrees, your opinion will prevail.

Or, suppose a reviewer says the statistical procedure you used was inappropriate. If you can demonstrate to the editor that the method you used was at least comparable, if not superior to the one the reviewer recommended, then there is no question with whom the editor would side.

Here to help

At least a few undergraduates have the notion that there is someone in their university whose sole job is to make certain nobody ever graduates. Many authors have a similar notion; that the main function of editors and reviewers is to find enough flaws to justify the rejection of every paper sent their way.

If an occasional reviewer or editor ever had that notion, it was never ASA–CSSA–SSSA policy. The function of editors and reviewers is to help authors publish the best possible paper. One scholar has described this process as "Protect readers from authors and authors from themselves." I put it humorously: "We're here to make you look good."

Don't be intimidated by editors and reviewers. After all, they are (in the words of the cartoon character Charlie Brown) sort of human, too. Chances are good, in fact, that you have been a reviewer yourself, and perhaps even served on an editorial board or two.

Reviewers are probably right most of the time. I have often read a sentence like the following in an acknowledgment: "thanks to an anonymous review for brilliant insight and valuable contributions." Of course, I have also heard complaints about reviewers who were totally wrong about an important scientific fact.

You must respond

If you fail to respond to a reviewer's comment or question, the editor may incorrectly conclude that you overlooked it, and ask you about it again. If that happens, you might mistakenly conclude that the editor is forcing you to comply with the reviewer's suggestion. Misunderstandings such as this may have contributed to the notion that reviewer comments must be followed.

While it is true that authors must respond to each reviewer comment, that response does not have to be positive. In fact, your obligation to the scientific integrity of your paper requires that you refuse to go along with wrong or wrong-headed ideas. In cases like that, heed the advice of Mrs. Reagan and "just say no."

* * *

I reiterated those thoughts again, in this column from March 1998:

Reviewer suggestions are just that

I write and speak two lines so often that I am almost embarrassed to repeat them, they have become clichés to me. Yet, every now and then I receive a comment that leads me to realize not everyone has heard or read them, so I will state them again here:

1. Reviewer suggestions are just that, *suggestions*.
2. Test everything and hold onto what is good. (This line is a quotation from St. Paul in the Bible, 1 Thess 5:21.)

What made me realize I have not repeated these phrases often enough was a recent letter from an author. The letter began, "I am revising a manuscript for [a journal] in which a reviewer suggested I add prepositional phrases . . . . I would like your comments about how the phrases add to or detract from sentence content. I am trying to improve my writing style without weakening it with extra words."

The author then quoted the original and suggested revisions of two sentences. Here is the original version of the first: "By using discriminant analysis to evaluate percentage clay, hue, and clast weathering in B horizons, he separated 135 footslope deposits into young, intermediate, and old colluvium."

The reviewer wanted the author to reword it this way: "By using discriminant analysis to evaluate percentage clay, hue, and weathering of clasts in B horizons, he separated 135 deposits on footslopes into young, intermediate, and old colluvium."

After I read those two sentences I concluded that the author had no reason to "hold onto" this suggestion from the reviewer, it made the sentence far more difficult to read.

To specifically answer the author's question, though, I don't worry so much about whether to use a prepositional phrase or any other particular structural technique. Rather, I tend to be pragmatic and use whatever works in a particular situation. Generally speaking, though, what works best is the simplest possible construction in each case. I believe this is especially important when the sentence itself presents complicated information, as this author's does.

There is a way to improve this author's already good sentence, though; reverse the front and back halves of it: "He separated 135 footslope deposits into young, intermediate, and old colluvium by using discriminant analysis to evaluate percentage clay, hue, and clast weathering in B horizons."

Aside from switching the halves, this is a word-for-word copy of the author's original sentence. Why does something as simple as switching the front and back clarify a sentence so dramatically? It doesn't always, as I will show next, but this time it created a declarative sentence; subject, verb, object. This type of sentence normally is the easiest-to read format, whether the sentence is "John runs fast," or "He separated 135 footslope deposits into young, intermediate, and old colluvium by using discriminant analysis to evaluate percentage clay, hue, and clast weathering in B horizons."

Here's what happens when the reviewer's version is turned around: "He separated 135 deposits on footslopes into young, intermediate and old colluvium by using discriminant analysis to evaluate percentage clay, hue, and weathering of clasts in B horizons." This sentence is nearly opaque, giving further evidence that the author's version was better.

The second sentence the author asked me about was already a well-written declarative sentence: "Relatively sharp Munsel color contrasts and abrupt textural changes separate two depositional units in the soils."

The reviewer wanted prepositional phrases added so it would read: "Relatively sharp contrasts in Munsel color and abrupt changes in texture separate two depositional units in the soils." This is another suggestion the author would be best not to heed because these changes only make the sentence more complicated.

Though the author's sentence is already well written, I recommended a way to improve it by using, of all things, passive voice. That I would recommend that construction in this case after so often writing against it is evidence of the pragmatism I referred to earlier.

My suggestion also turns the author's sentence around: "The two depositional units in the soils are separated by relatively sharp Munsel color contrasts and abrupt textural changes."

In this sentence passive voice is (slightly) preferable to a declarative sentence because the original sentence's subject was long and complicated: "Relatively sharp contrasts in Munsel color and abrupt changes in texture." That is a lot of syllables to wade through before finally getting to a verb. The sentence's object, "two depositional units in the soils," is shorter. Placing it first allows the reader to get to the verb more quickly; reasonable trade-off for the weakness of the verb, "are separated."

I hope the author did not "hold onto" my suggested revision of that second sentence before testing it. My version implies *only* two depositional units, which may not be the case in the plot under discussion.

The author showed some understanding of St. Paul's rule when he wrote: "I . . . do not believe [the reviewer's suggestions] help the first sentence . . . , [but] many other editorial suggestions made by the reviewer improved my manuscript."

That reminds me of an incident involving Abraham Lincoln. He asked 12 of his most trusted advisers to help him decide a complicated matter. All 12 voted that the President act in a particular way.

"I guess that's what it will be, Mr. President," one of them remarked.

"Not necessarily," Mr. Lincoln replied. "I haven't cast my vote yet."

Whose paper is it?

When it comes to dealing with reviewer or editor suggestions, the most important thing to remember is whose name appears under the title of the manuscript. If your name is there, you are the person who is going to have to defend that paper to your peers and critics. It seems to me that any author would find that difficult to do unless he or she honestly believed the paper gave the best possible presentation of the information.

Read and carefully consider every suggestion from every reviewer. I often read in a paper's acknowledgment a sentence along the lines of, "Thanks to an anonymous reviewer for thoughtful insights and excellent suggestions." By all means, "hold onto" such insights and suggestions.

If there is a rule of thumb regarding what to hold onto, it would be to consider the reviewers, technical editors, and associate editors as more expert on scientific and technical matters; and the professional editing staff employed by the publisher as more expert on style and grammar.

* * *

One step taken by some journals to make reviews less biased and more fair is to keep author anonymous from reviewers as well as vice-versa. When two ASA–CSSA–SSSA journals decided to take that step, I wrote the following column to explain the rationale behind it. The column was published in May 1993:

Why some journals use double-blind reviews

Several authors have asked us why a couple of our journals are experimenting with double-blind reviews. The answer is fairly simple, a majority of surveyed members asked us to. A 1990 membership survey revealed a concern that biases might cause reviewers to be too hard on some authors and too easy on others. Survey respondents asked the

Societies to explore the possibility of having authors anonymous from reviewers as well as vice-versa. (In that same survey, a sizeable minority asked that both authors and reviewers names be revealed. We will not likely try that any time soon.)

[That previous parenthetical statement was all I thought necessary at the time I wrote the column being presented here. The question continued to come up subsequently, so I will insert the reason why this suggestion is not a good idea. Basically, it is a quick and easy way to cheapen reviews. Some reviewers are afraid to give a bad review to someone they fear. Others will deliberately give a gushing review to someone they are trying to impress or perhaps for someone they hope to gain employment from. Every time double-open reviews have been tried by any journal they quickly have been abandoned.]

All editorial boards were asked to consider trying double-blind reviews. At the 1991 annual meetings, *Agronomy Journal* and the *SSSA Journal* editorial boards indicated that they would try it if their respective Societies' Boards of Directors wanted them to. The Boards told the journals to proceed with two-year tests. (The minutes of these meetings were published in the March-April 1992 issues of those journals.)

Biased reviews

Readers and reviewers can be prejudiced for or against a piece of writing based on what they know about the author. This was demonstrated in at least two studies conducted about 30 years ago.

One of the studies involved scientific literature. An accepted paper, before it was published in a journal, was retyped in two versions as though it were a new manuscript. The only difference between the two versions of the paper was that one manuscript was given the fictitious byline, "By Dr. John Smith," and the second, "By Dr. Joan Smith."

These manuscripts were then sent to a group of reviewers normally used by the journal, but who had not seen the paper. The reviewers unanimously believed that "Dr. John Smith's" manuscript was an excellent addition to the scientific literature and accepted it without hesitation, making few, if any suggestions for improvement. The reviewers who read "Dr. Joan Smith's" manuscript also accepted it (most of them did), but not nearly so readily and with many more requests for additions, corrections, and modifications. It would be interesting to see what would happen if the same sort of test were conducted today.

In another test, 50 quotations, all by Benjamin Disraeli, were assembled, divided into two groups, and printed on a test instrument. Below one group of 25 quotes the researchers printed false names, attributing the words to persons such as George Washington, Abraham Lincoln, Winston Churchill, St. Francis of Assisi, and other persons generally recognized as "good." The second 25 quotations were attributed to such persons as Mao Tze Tung, Adolf Hitler, Joseph Stalin, Al Capone, and other persons generally recognized as "bad."

The cards on which these statements were printed were shown to a group of people who were asked, "Forget who said the words, do you agree with any of these statements?" The vast majority strongly agreed with statements supposedly uttered by the "good" people and mildly to strongly disagreed with statements attributed to the "bad" people.

My third example is anecdotal. While still in school, I read a letter to the editor of the local newspaper and thought it was one of the worst-written letters I had ever seen. It seemed shallow, trite, and lacking in substance. I noticed, when I got to the end, that it was written by my creative writing professor. I took another look at the letter and this time noticed many good things about it that I hadn't seen the first time.

Thus, prejudice for or against authors has been shown in the past, and continues at least somewhat today, and probably into the future.

Most scientific reviewers try to be careful, but with good intentions they can tend to be more kind with authors they like and more thorough with authors they either do not like or whom they believe to be less informed. "She (or he) is a student; I need to correct this so they won't make the same mistake the next time." "Old Harry (or Hortense) is an excellent scientist, but getting along in years. This statement isn't quite perfect, but it is close enough; I'll let it go."

Most scientists are aware of these occasional biases. More seasoned scientists (who benefit from it) tend to accept these biases as part of the game. Younger and minority scientists (who may be hurt by it) are less sanguine, especially when the biases are unfair to them.

Opposition to double-blind

I've noticed two sorts of opposition to double-blind reviews. One appears to be from people (primarily prestigious scientists) who *want* their name associated with the paper. There is no answer to this type of opposition; one cannot blame a person for wanting to make the most of a hard-earned and well-deserved reputation. The phenomenon is sometimes called "the halo effect." The idea it is that if the person is prestigious, all of that person's work will be also.

The other type of opposition to double-blind review seems to be from scientists who think it is not worth the effort. More than one person has said, "I can generally tell who wrote a paper just by noting the subject matter and location." This is easy to refute: Granted, the relatively small number of scientists doing the sort of work we publish makes it easy to identify the lead scientist on many research projects. There is no way to tell, though, whether that research leader was the senior author of a particular paper, or if its senior author was another full or associate professor, or a graduate student or visiting scientist.

Some reviewers have objected to double-blind reviews because they don't know whether to be hard or easy on a particular author. Others would argue that this is *precisely the reason for* double-blind reviews; papers are judged on their merit instead of the personality of the author.

How to prepare a paper for double-blind review

First of all, it is impossible to remove all clues to the identity of the author(s). You cannot falsely say that the work was done in Florida when it was actually done in Colorado or Illinois. If you are the only person who is working on a particular crop, you cannot leave out the name of the crop ("this research was conducted on anonymous plant material") in order to hide your name. Similarly, there may be a clue in the type of soil; but you cannot leave out or falsify the soil name to hide your identity. Neither is it necessary to delete all references to your own papers in the Reference section. Any of the above, if they were done, would obviously compromise the scientific integrity of the paper.

All the journals trying double-blind reviews ask is that you use a title page that can be removed from the rest of the paper. On it, list the article title, byline, author-paper documentation, and (if the paper has one) the Acknowledgment. Repeat the title, but none of the additional information, on the first page of the paper; the page that contains the Abstract.

Also, of course, do not use your name in the page numbers—use "Page 4," not "Luellen, page 4." Be careful how you identify tables and figures; key them to the title instead of to your name. In other words, instead of "Luellen, Fig. 2," use "Fine-tuning your writing, Fig. 2."

* * *

There is still one more level of review performed on scientific papers. That is the copyediting done by a professional editor hired by the publisher. This essential level of editing has caused confusion among many an author, as I outlined in this column from June 1993. Although this column deals specifically with editors hired by ASA–CSSA–SSSA, all publishers employ such editors either full time or on a free-lance basis.

Why Madison editors ask you to change an approved paper

According to an old joke, the three greatest lies are, "The check is in the mail." "You know I will always love you." and "I'm from the government; I'm here to help."

Judging by some of the comments we receive, a few of our members would add a fourth to that list; "I'm a Headquarters editor; I'm here to help you improve your paper."

I am probably biased, but that final statement is not a lie. A typical scientist will write perhaps 200 papers during his or her lifetime; a typical professional editor at Headquarters handles 200 papers *every year or two*. That fact alone indicates that we have some experience you can tap into.

The Societies employ 10 full- or part-time professional editors, a professional artist, plus support staff at their Headquarters. I have sensed that some authors believe these editors may know writing, but not science. That, also is not true. One has a degree in soil science, another a degree and laboratory experience in biology, a third was the in-house editor for a science department at a major university. And, of course, all have had experience dealing with agronomy, crops, and soils scientific papers.

As most of you believe, our editors also have extensive experience with others types of nonfiction writing. One edits and, with her husband, publishes their own community newspaper; a couple other of our editors work as part-time reporters for their community newspapers; still others work part-time as free-lance authors of magazine articles.

The job of Headquarters editors

There seems to be some misunderstanding among Society members as to where these professional editors fit into the scheme of things. This confusion comes about, in part, when an author hears from a member of an editorial board that the paper has been accepted for publication in the journal; then, a few weeks later, gets a letter from Madison with one to 30 questions asking for further clarification of the text, improvements to a figure, and other changes before that paper can be published.

"What is going on?" The confused author asks. "Was the paper approved or not? Is it going to get published or not?"

Yes, the paper was approved by scientific editors; and yes, it will be published as soon as all style and format questions are dealt with.

In short, there is a division of labor. The editorial board is responsible for the scientific accuracy of a paper, and the professional editing staff is responsible for making sure that a paper adheres to style guidelines, and helping ensure the material is presented well.

Origin of our editorial style

Have you ever wondered why Headquarters editors ask you to identify manufacturers of commercial products the first time they are mentioned; to give the binomial scientific name for species after their first mention in abstract or text; to use chemical symbols instead of words for elements, ions, and compounds (except at the beginning of a sentence); and all the seemingly nit-picky details that they say need to be cleared up on a paper you thought was completed and accepted for publication?

The ASA–CSSA–SSSA editorial style was established by the Societies, not by the Headquarters staff editors. The ultimate authority over editorial style is the *ASA-CSSA-SSSA Publications Handbook and Style Manual*. The ultimate authority over this book is the Boards of Directors of the Societies. Any style changes must be approved by all three of these Boards before they can be implemented and placed into the style manual.

The Boards of Directors delegate the responsibility for maintaining the manual to the Editorial Policy Coordinating Committee (ACS321). This committee is composed of the editors-in-chief of the Societies, the editors and managing editors of each journal, the chairs of the ASA monographs and the SSSA book series committees, and the executive vice president and the associate executive vice president of the Societies.

Changes in style are usually suggested at one or more of the journals' editorial boards, discussed by editorial boards of other journals, and by the Editorial Policy Coordinating Committee. If this committee agrees to the proposed change, it is taken to the Boards of Directors for final action.

Division of responsibility

Some authors have responded to our queries with a question along the line of, "The professional scientists who reviewed this paper weren't bothered by that during the review process; why should I change it now?"

That is a good question. The answer is that style and grammar questions are not the primary responsibility of editorial board members.

They will often ask style and grammar questions, but there are at least three reasons why they may not:

- An editorial board member takes the division of labor seriously and concentrates on scientific questions, leaving all style matters and language questions to Headquarters editors.
- An editorial board member will ask correction of style and grammar errors in the specific parts of the paper that also had scientific questions, but leave the rest of the paper to Headquarters editors.
- An editorial board member may have asked for some style and grammar corrections, but inadvertently missed others.

As well as clearing up style and grammar errors, the professional editing staff at Headquarters is also encouraged to make suggestions to authors on how they might improve the readability of their papers. Style and grammar errors must be corrected; suggestions for improving readability are just that—suggestions.

More about Headquarters editors

As well as their writing and editing experience, Headquarters editors have a great deal of experience with publishing tables and figures. They can tell, for instance, whether a figure will look all right or if some of the lines on it will be too light or the numbers on it will be too small to reproduce clearly. They know how to get the most possible clarity out of a murky photograph. You can argue with a Headquarters editor about this —and, as with writing suggestions, you generally will get it your way— but you may also get subsequent complaints from colleagues who couldn't see the lines or read the numbers in your figures.

After reading all of this, you may still grumble when we ask you to make changes in a paper you thought was ready to publish. (I sometimes grumble when an editor asks me to make changes in a paper I thought I had completed writing.) But as you grumble, please realize that (i) we are only following the style manual and doing the job the Boards of Directors has hired us to do, and (ii) we honestly believe, based on our knowledge and experience, that the changes we suggest will improve your paper.

Chapter 16

Electronics and electronic publishing

I believe I may have been the first person to present information in a periodical about writing that outlined problems with the electronics that underlie the words written on a computer. Judging by the mail response this article generated, others have noticed something of the problem, too. Here is that article, as presented in the February 2000 issue of my newsletter:

Think about electronics, not just looks of a paper

Most journals nowadays are edited electronically, and more and more journals are published electronically. What difference does that make to authors? Simply this, authors need to begin thinking as much about the underlying electronics of their manuscripts as they think about the way their manuscripts look on paper. In fact, before much longer, the underlying electronics of how a paper is set up will be more important than the physical appearance of the hard copy of that manuscript. You read me correctly—not just as important, *more* important.

The reason is because machines are taking over more and more of the final steps toward publishing. These machines read electronic pulses, not marks on a piece of paper. Unless drastic changes are made in hardware and software in the next couple years, that simple fact will continue.

Today for most journals

If you are writing for a conventional journal, such as most of those published by ASA–CSSA–SSSA, you are probably aware that the professional editing staff in Madison uses a combination of visual and electronic editing. They've edited the *text* of papers electronically for at least a half-dozen years, and those texts have been typeset electronically for a longer time than that.

A hard copy of the text of each paper is still sent to the typesetting plant along with the edited disk copy of the manuscript, but that paper copy (at least as far as the text of the paper is concerned) is used primarily as a guide and a back-up to verify the electronic copy, which actually drives the typesetting equipment.

For ASA–CSSA–SSSA, the equations within the text and tables that accompany the text are still edited by hand and typeset mainly by hand. As for figures, no one method predominates. A few figures for a few publications are still shot with a camera from hard copy. Most nowadays are scanned electronically from hard copy, and an increasing number are retrieved electronically from a disk or other electronic submission.

Electronic journals

One of my major clients is the National Cotton Council of America. I work with the people there on their *Journal of Cotton Science.* This peer-reviewed journal is published only on the Internet. It has never published paper copies and probably never will.

So all edited manuscripts I send to that journal are electronic, not on paper, and must be ready electronically, as well as grammatically and stylistically. I do not have the luxury of sending a paper copy for someone else to scan or keyboard.

Author input

Most authors, I have found, do a reasonably good job of preparing the electronic version of the *text* of their papers, regardless of the journal they write for. And, interestingly, the equations supplied by authors with those papers are generally in pretty good shape. So too, for the most part, are figures. But three areas often present me horrendous problems.

Unnecessary codes

If, as you write a manuscript, you copy something from one document to another you may unwittingly bring along some superfluous codes with it. Either that, or the computer itself will insert some codes without your knowing it.

For example, if you copied a reference listing from a previous paper you had written, and that earlier paper was written in a different typeface than you use now, that material probably copied in its original typeface. If those two typefaces were similar, you may not have noticed the difference. You may have also brought along some line spacing, word spacing, type size, tabular indicators, and other codes. You may not notice these unneeded codes unless you specifically looked for them.

Further, when you delete something you may leave behind some or even all of its electronic coding. For instance, if you delete or move a word that was italicized (such as a botanical name), you may leave behind a pair of "italic on—italic off" codes with nothing between.

By the time a manuscript is completed it can have a fairly large quantity of additional and unneeded codes for line spacing, hard indents, hard tabs, line advances, widow and orphan protection, and many others items. A large portion of these codes will not be apparent on either the paper printout or the on-screen version of the paper. For instance, a blank space at the end of the last line of a paragraph could contain any number of hard tab and hard indent, italic on—italic off, and bold on—bold off codes. You will likely never get any indication of any of those unless you specifically look for them with the *reveal codes* feature. Unfortunately, MS Word software does not have such a feature.

If you use WordPerfect, though, I urge you (if you don't already) to open the *reveal codes* every now and then and delete any extraneous and redundant codes you may find.

If you've never done this, you might be amazed the first time you do it at the number of extraneous and redundant codes you find. In some manuscripts I've handled I've seen an average of one or more per sentence. Some particularly messy manuscripts have contained a dozen or more (on rare occasions, 50 or more) per sentence. (Tables, as I'll mention below, can contain far more than this.) It has, on occasion, taken me a full hour to delete all those unnecessary and redundant codes just from the text before I could begin to actually edit it—and this represents time spent *after* a macro has deleted all the codes that can be removed without the input of editorial judgement.

While unnecessary and redundant codes are a nuisance, they do no actual harm to a manuscript—until it is ready to be published electronically. Those mentioned in the next section can lead to costly problems.

Incorrect codes

Some automatic codes that come in handy for authors provide a costly nuisance to editors and publishers. Automatic outlining is one such code. Simply by hitting the "return" or "enter" key after one item in a numbered list, the computer automatically inserts the correct indenting and numbering for the next item in that list. This makes setting up an outline easy for an author creating a copy for a paper printout.

Typesetting and Web-publishing machinery might be able to interpret those codes for the first line of that text, but do not understand and follow it from then on, giving the machinery the electronic equivalent of a nervous breakdown.

Professional editors need to delete all the automatic outlining and insert actual numbers, spaces, indents, and so forth so as to create the kind of electronic pulses the machinery will be able to read.

Automatic footnoting causes similar problems. Typesetting machinery is trying to set type for the format programmed into it. Often these machines cannot see the imbedded footnotes at all. And those machines that see it do not know what to do with it.

Professional editors must find those imbedded footnotes (not always an easy task) copy them to their correct place at the beginning of the manuscript, and then delete the imbedded note.

Another nifty feature of MS Word is its "first line indent." Instead of a hard tab at the beginning of each paragraph, this automatically indents each line after a hard return. No doubt authors love it (I don't it gives me no flexibility as I write). But again, the machinery reads the first one and then either sets the rest of the text as a single paragraph or has the breaks after each paragraph, but fails to indent the first line of the next.

So, someone like me needs to remove that single code at the beginning and insert a tab at the beginning of each paragraph throughout the manuscript. The problem with this if I delete the code before inserting its replacements, it's not always easy to find the beginning of each paragraph. For paragraphs that end with a short line, this search is not too difficult. But I have a more difficult time finding the end of those paragraphs that end with a full line. Lately I've been using the "search and replace" tool to find the "hard returns." I replace that single hard return with a hard return and a hard tab. Before I learned to do that my search for the beginning of each paragraph could be extremely tedious and time consuming.

Reference sections

As you know, reference sections are generally set up in the opposite format of a standard paragraph; the first line of each entry is on the margin and all subsequent lines of that entry are indented. In printer–editor jargon, this format is called a *hanging indent.* Hanging indents are extremely easy to set up in most word processors. In most WordPerfect releases, for instance, with the curser at the beginning of the first line, pressing "control" and "F7" does it instantly. In MS Word, place the curser at the beginning of each entry and, with the mouse, go to "format, paragraph, indentation, special," and finally "hanging indent" and click the left button on the mouse.

I wish all authors knew the information in the previous paragraph. And I further wish all authors who do not know that information (or would rather not go through all those steps with MS Word) would not try to set up their reference sections with their own coding pattern to create hanging indents that look good on paper but are a jumbled mess of electronic codes underneath.

When I set out to clean up an author-created set of hanging indents in a reference list I first need to figure out how the author created the list. Usually when I open the "reveal codes" I find some combination of hard returns and indents or hard tabs, with an occasional "first line indent" or "hard advance" or something else thrown in for good measure.

One combination I see probably more often than others involves a hard return at the end of each line in a reference entry and a hard tab or a hard indent at the beginning of each line after the first. This adds up to five to seven codes per entry, invariably accompanied by several extra hard spaces thrown in for good measure.

My first task, or that of any professional editor, is to strip out all those incorrect codes from each reference entry and then insert the single correct *hanging indent* code. Depending on how many codes need to be stripped, this process might take 5, 10, 20, or even 30 seconds per entry. On a reference list with perhaps 30 entries this could total 15 minutes.

Remember my comment at the beginning that underlying electronics are more important than the appearance on paper? If the author had set each reference section entry as a standard paragraph it would not look right on paper, but it would take me less than half a second to delete the single *standard indent* code and replace it with a single hanging *indent code.* I could correct the same 30-item reference list mentioned above in perhaps 15 seconds, and less than that if I set up a global search and replace to correct it.

Going electronic with tables

While reference sections can present a professional editor or typesetter some minor headaches, the underlying electronics with tables can present major industrial-strength headaches and possibly even accompanying nightmares.

I have seen a thing that looked like a single table in the hard copy that was actually three or four or five electronic tables stacked on top of each other. Still others might have been created as a single table, but have dozens of lines of text in each cell instead of only one. Still other tables have items that should have been placed into a single cell spread across two or three or more cells.

Other things I've seen looked like tables on the screen and on paper, but were created with the computer's *newspaper column* formatting tools. These tables have a stack of items in one column, spaced as necessary vertically to keep them lined up as they should be. If the author does a good job of vertical spacing in each column, the "table" does not look bad on the screen or paper printout. The problems come when an editor or typesetter tries to convert this electronic mishmash into a real table. Keeping the various parts of each row lined up with each other across the table can be, to say the least, difficult.

In Chapter 10 of this book I describe and demonstrate the only two acceptable tables, using either the table-creating feature of the software or with single tabs per line and varying as much as needed the information concerning those tabs.

Unfortunately, many authors do not do this. Instead, they line up their columns of items visually with an eclectic combination of one to five tab stops plus a varying number hard spaces between each item in each line. Again, this looks fine on the computer screen and paper. And it is easy for anyone to keyboard such a table from the paper copy (which is the way it is done at ASA–CSSA–SSSA). The professional editors there simply delete the electronic copy of the table from the disk they send to the typesetter and have the typesetter re-key the table—with the concomitant cost and risk of typographical errors.

The machinery that typesets electronically or publishes a table on the Web must have only a single electronic pulse between each item in each row. When I edit for an electronic journal such as the *Journal of Cotton Science* I do not have the luxury of deleting this electronic mishmash and asking some typesetter down the line to set the table for me. I must either manipulate the electronic copy in front of me to construct a new viable table from the electronic mishmash supplied by the author or re-keyboard much, if not all, of the table myself.

The more of these tables I work on, the better I am getting at manipulating them. Sometimes I am able to properly set up an incorrect a table in less than an hour. This, of course, means I get to bill for an extra hour of work per table, which I appreciate, but the client does not.

Gearing up for change

What difference do the problems I mentioned in this column make to you? Why should you have to learn a new way of doing things just to make my life easier? "After all," you might say, "I *am* getting paid to clean up after you."

True, I am getting paid, quite well, in fact. And the more work you make for me (and other professional editors like me), the more I am able to charge for getting your paper ready to publish. While this is fine by me, it is not so good for you, as mentioned two paragraphs ago, because it drives up your publisher's overhead.

The impact will be almost immediately apparent to you when you publish in journals that assess a page charge. The impact will not be so immediately visible to you from those journals that recoup their costs in some other way than page charges. For these journals, it may contribute toward a lowering of the acceptance rate so the journal does not publish quite so many papers, or maybe a raising of the subscription price the journal charges libraries. Both the above can directly affect you in terms of either fewer of your papers published or fewer journals available in your local library. If the journal is published by a professional society that does not have page charges it can increase, problems of the sort mentioned in this paper could be one of the contributors toward your next dues increase.

In short, it is to your advantage to not only learn more about the underlying electronics of the papers you write, but more importantly, to start writing your papers so they look just as good electronically as they do on paper.

How soon should you begin to learn these keyboarding and other skills? It depends to a great extent on the publication you are writing for. But even if you write for a conventional journal, such as most of those published by ASA–CSSA–SSSA, it would pay you to begin to learn how to do it correctly. You will not only be placing yourself a little ahead of the need curve, you will be ready for the additional changes that are quickly coming down the line, and on line.

Chapter 17

Poster presentations

The contents of this chapter, fittingly, is the most recent I have written. It comes from the January 2001 issue of my newsletter, which was not mailed for more than a month after these words were written.

Move your poster presentations to a whole new level

Strolling recently through a large poster hall at an international meeting of scientists I couldn't help noting that virtually every poster I saw had at least one major flaw in its presentation. Then a further thought occurred to me, I cannot remember ever reading anything substantial on the presentation of posters. Sure, there are a lot of beginning-level pieces around, but I've never seen any material at what might be called the graduate or professional level. Therefore, necessity being the mother of invention, that is what I intend to present in these pages.

Posters have come a long way during the past couple decades. Those old enough to remember that far back may recall an occasional poster or two that were little more than simply pinning up the manuscript pages of a paper. There have been a great many changes in posters in recent years, brought about in part by the development of improved typesetting quality of word-processing systems and improved output of printers. But if you

---

look closely at some of the posters still being presented today, the contents of far too many is still closer to that of a journal paper than to a poster.

The notes I jotted down as I walked through that poster hall include:

- not using anywhere near the space available
- too much type
- too small type
- unreadable type font
- lines of type too long
- references cited within a poster, but no reference list posted
- murky photos
- poster-on-picture; risky, can be done, but often terrible
- type on solid or graduated-color backgrounds can work, but watch colors and contrasts

I will deal with all the above, and more. Perhaps a good place to begin—because it will help solve many of the other problems—is the simple fact that I did not find a single poster that took advantage of the space available for it.

Size of presentation

Most posters I saw at this most recent meeting fell into one of two categories.

First was what might be called the traditional way of posting material. This involves the pinning or otherwise attaching individual panels to the poster board, to create a display that looks something like this:

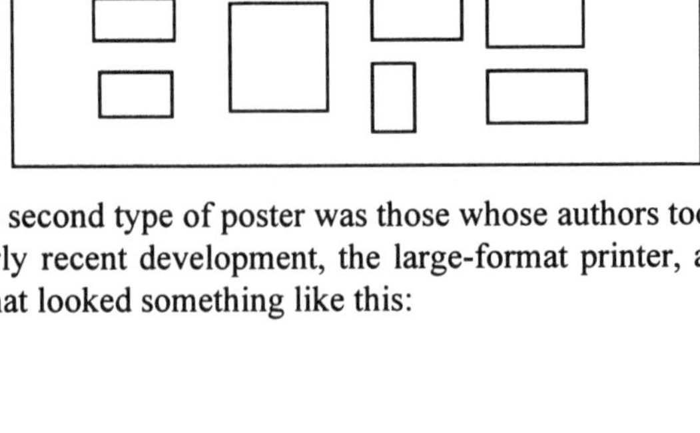

The second type of poster was those whose authors took advantage of a fairly recent development, the large-format printer, and created a poster that looked something like this:

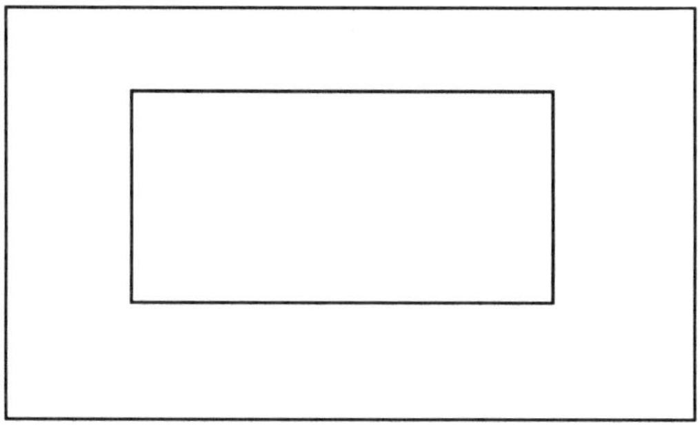

Do you see something common to both of these setups? In each case, they occupy only about half (or less) of the space available on the board.

Is there a better way? Quite simply, yes. The thought occurred to me almost immediately. When I suggested it to several scientists I could almost see the clichéd knock to the forehead and hear the words, "Of course! Why didn't I think of that?"

What was this magical solution? Place your type on one of those large-format poster printers in portrait instead of landscape configuration and use more than one sheet per poster board to create something that looks more like this:

This configuration gives you a minimum of twice as much space to work with. I don't want to find myself in the position of Dr. Frankenstein, though, so I hastily add the hope you will take advantage of this additional space to create a better poster presentation, rather than simply to shovel more words into that space. After all, the purpose of the poster is to present the highlights of your work. The details are to be filled in during your conversations with those asking their questions.

What steps can you do to create a better poster? Five of my nine random notes above refer to type, so let's begin there.

Size, color, and contrast of type

Most type I saw on most poster boards was too small. Some of the type I saw that was of about large enough, but was not as legible as it might have been. Back in our days at ASA–CSSA–SSSA we had a surefire way to tell if type was readable. We would take a sample of it to the person in the office with the thickest glasses. If he or she could read it, we figured most everyone else would be able to read it also.

You might try something like that with any poster you are developing. The question is not whether *you* can read it, standing at the distance away from it you likely would be standing in a poster hall. The question is whether *anyone else* can read it. Ask several people, with varying visual acuity whether they can read it. You don't necessarily have to prepare a poster that is readable by the legally blind, but try to make it readable to a representative cross-section of the segment of humanity most likely to be present in the poster hall where the paper will be presented.

Use of colors of type different from black is neither good nor bad, it depends on the colors of both that type and the background. I saw one poster that had gold type on a dark blue background that looked simple, elegant, and extremely readable. I saw another poster with small green type on a murky purple background that was all but illegible. Choose your color combinations for both readability and attractiveness. I mean that final word both in its common meaning and also in its best sense—something that attracts readers to it.

Regardless of whether you choose black type on white poster boards or some other combination, the first thing to keep in mind is readability. And the first component of that is enough of a contrast so the type can be read. It is no mistake that the signs you see along Interstate highways are either white on green or white on blue. Many hours and many research dollars went into a study of the best and most legible color combinations before those colors were decided upon.

I'll have more about backgrounds below, but first I want to get to another important matter, the length of your lines of type.

Length of lines

If you have ever looked at the original or accurate reproductions of the Declaration of Independence, the Constitution, and other early documents you will see that they are composed of exceptionally long lines of exceptionally small type. That combination makes them nearly unreadable without laying a straightedge or piece of paper on them, or at least moving your finger down the left side to help you find the beginning of the correct next line.

By contrast, if you have ever seen type alongside a photo in a newspaper you have seen lines so short that words don't fit well within them. The result is either large gaps between a couple small words or tons of hyphens dividing long words.

Just as there have been studies of color combinations, mentioned above, there have been studies of the readability of line lengths. Those studies determined that the most readable line length is between one and two alphabets long, with 1.5 alphabet lengths being the optimum.

In other words, regardless of the size of type you use in the text blocks on your poster, shoot for an average of somewhere between 26 and 52 letters per line, with 39 as your optimum target.

At that scientific meeting referred to at the beginning of this paper I saw several text blocks with lines that contained more than 50 letters. One block of type contained close to 100 letters per line and another had somewhere between 150 and 200. That latter one I didn't count precisely because doing so might have embarrassed the author.

As mentioned above, the only way some extremely long-lined blocks of text can be read is by laying a paper or straightedge on the text to help yourself find the next line. It would be both difficult and embarrassing to do that in a crowded poster hall.

The remedies for too many letters in a line of type are simple; enlarge the type size, shorten the line length, or some combination of both.

Clarity of backgrounds

As promised in the section about type contrasts, I want to bring up the subject of the various sorts of backgrounds I found under the type on many posters. Backgrounds can be a wonderful idea, a good idea, or a terrible idea, depending on the type and the background.

Further, if you use a solid color, at least consider coordinating it with the meaning of your poster. One poster on water-use efficiency was printed on a pink background. Another, on chlorophyl, was printed on a brown background. Neither of those made any sense to me, but at least they did not detract from the subject. One poster on cleaning up water appeared on a brown background and carried exactly the opposite message of the material in the poster printed on it.

Now a poster on soil would make a great deal of sense on a brown background, and one on some type of blooming flower or perhaps on a litmus test might be quite logical on a pink background. For cleaning up water, a blue or perhaps a blue–white gradient would make much more sense than the use of muddy brown.

Keep contrast in mind, though, regardless of background you use. I saw two posters, both with white type [which I'll have more to say about in a moment] that were on pale backgrounds—one light blue, the other, gray. The type was barely legible, let alone readable.

I also saw one with black type on pale green that looked quite nice.

Backgrounds with graduated color or colors can look nice also, so long as you are aware of the contrast across the entire poster. The following figure demonstrates this principle.

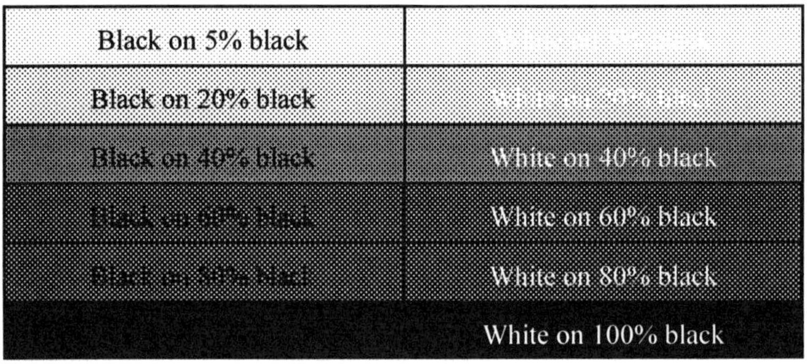

Many of the posters on the large single sheets of paper were superimposed on a photo. This can be a good or a terrible idea, depending on not only the contrast, but also the material in the background.

One poster had a large photo of a corn field. The author's tables, already printed in far-too-small pale blue type, appeared among the tassels of the corn and were barely legible.

Another poster of a turfgrass experiment was superimposed on a large photo of a soccer field. The premise made sense, but unfortunately that particular field had a billboard attached to the far fence in exactly the correct position so that the far-too-prominent words "Fuji Film" appeared directly beneath the introduction. It was incredibly distracting.

Solid color or graded color backgrounds can work wonderfully. They get around the problems that can happen with photos. But keep in mind the colors you are using for both the background and the type.

Reversing type out of a background

I have referred to legibility—both through the demonstration above of black on black and white on black—and by mentioning that the white on green and white on blue of the Interstate highways are highly legible. I also mentioned that two posters I saw had white on pale blue and white on gray and were not very legible. The printer jargon for lighter-color type on a darker background is *reversing* the type out of the color. Type reversed out of the color can be extremely striking when handled correctly and, as the Interstate highway signs attest, can be highly readable. But there are some legibility considerations to keep in mind.

The main rule of thumb to remember is that any reversed type will not be as legible as its nonreversed counterpart, so must be presented larger and/or bolder than normal. The type in the following two boxes is exactly the same 12-point Times New Roman. As you can see, the type on the right looks smaller and fainter.

Black on white	White on black

In the following example the type on the left remains the same 12 points, but the type on the right was enlarged to 14 points.

Black on white	White on black

Both sets of type in the next example are again 12-point Times New Roman, but the type on the right is bolface, which also helps its legibility.

Black on white	**White on black**

Using a bold sans-serif type helps even more. The type on the left is standard 12-point Univers, that on the right is 12-point Univers bold.

Black on white	**White on black**

I wish I could give a simple rule of thumb regarding what to do to make reversed type legible—simply increase its size by a certain percentage or whatever—but so much depends on the type involved.

The thing to keep in mind is that reversed type *will not* be as legible, and you need to take some steps to help improve its readability. The precise steps depend on the type face used, the colors used, and a few other things as well. The best advice I can give is to play around with it and use whatever technique works best in your particular situation.

Because of the legibility problems, though, I would recommend that you use reversed type only for special highlights, not for the entire text of your poster.

Some tables, especially if they are not too complicated, can look striking. Bar graphs can really stand out when reversed. Various colors of bars on a black or maroon or deep blue background can look great.

Sans serif vs. serif type

In the poster hall mentioned throughout this paper I got into a long discussion with a scientist about the pros and cons of sans-serif type. He said he preferred sans-serif type for every part of a poster. I agreed with him that such type works well in tables and figures and is commonly seen in titles, by-lines, and other such places.

This is a typeface with serifs	This is a typeface without serifs

We disagreed on the subject of sans-serif type for text blocks, though, so we walked through the hall together looking at a representative sample of posters. After showing him several posters with the more traditional serif type, I pointed to one that had sans-serif type throughout and said to him, "It's cleaner, sharper, I agree, but I also think blocks of type of this sort are more difficult to read."

He looked at it for a few moments, took another look at a nearby poster that had serifs on its type, and said, "I think you might be right."

Then I told him something he had not been aware of; scientific studies on type readability have determined the same thing. He said, "Well, then. Maybe I'll try it on my next poster."

Tables, photos, and other figures

It is difficult to generalize about what I saw in this category. Some of the material was wonderful, some was marginal, and some was terrible. If there is one comment, though, it is that most photos are not large enough. One poster, for example, had eight photos across the bottom, each about the size of a small postal card.

It is possible that all of those photos were needed. Were I asked for advice, though, I would have asked the author to consider using about half as many photos, so those that remained could be presented in a larger size.

I believe color photos are probably essential on posters nowadays. Apparently authors believe this also, because I don't think I saw a single poster with black-and-white photos.

Most scientific journals tend to be stuffy about the colorful presentation of graphs. You can not only get away with more colorful presentations in posters, such treatment should be encouraged. When you are in a situation were several hundred scientists are walking among a forest of close to a thousand posters, anything that visually says, "Hey, stop and take a look at me!" is to your advantage.

Conclusion

There is no one best way to write a poster, merely a set of premises.

It should have a minimum amount of text. People standing on a concrete floor staring at a poster cannot physically read a great deal of textual material.

It should be physically attractive. Not necessarily pretty, perhaps, but approachable enough to make a person want to stand and look at it.

It should have something about it to catch the eye of people walking past it in a crowded hall.

It should be easy to read and easy to follow from one item to the next.

It should be understandable in a single reading. The crowds of people milling about the poster hall present a situation where few people will be able to reread complicated passages.

Index

Bibliography

For your convenience, all the books cited and referenced anywhere in the text are repeated here

The ACS Style Guide. A Manual for Authors and Editors—*Edited by Janet S. Dodd. American Chemical Society, 1155 16th Street, NW, Washington, DC 20036. 1986. 264 p. ISBN: 0-8412-0943-X.*

American usage and style: The consensus—*By Roy H. Copperud. 1970. Van Nostrand Reinhold, New York City.*

The Careful Writer: A modern guide to English usage—*By Theodore M. Bernstein. 1984. Atheneum.*

The Chicago Manual of Style : The Essential Guide for Writers, Editors and Publishers, 14th ed. *1993. University of Chicago Press, Chicago. 921pp. ISBN: 0-22610-389-7.*

Dictionary of Modern English Usage—*By W.H. Fowler and Sir. Ernest Gowers. 1965. Oxford University Press, New York and Oxford.*

Dos, Don'ts & Maybes of English Usage—*By Theodore M. Bernstein. 1997. Random House, New York City.*

Editors on Editing—*Edited by Gerald Gross. Grove Press, 196 W. Houston Street, New York, NY 10014. 1993.*

Effective Writing: for Engineers, Managers, Scientists—*By H.J. Tichy with Sylvia Fourdrinier. 1988. John Wiley & Sons., Inc. ISBN: 0-47180-708-7*

Effective Writing Strategies for Engineers and Scientists—*By Donald C. Woolston, Patricia A. Robinson, and Gisela Kutzbach. Lewis Publishers, 121 South Main Street, Chelsea, MI 48118. 1988. 181 p. Hardcover. ISBN: 0-87371-099-1*

The Elements of Style, 4th ed.—*By William Strunk E. B. White. 1999. 4th ed., 105 p. Macmillan Publishing Co., Inc. ISBN: 0-20530-902-X.*

GPO Style Manual. *U.S. Government Printing Office, Washington, DC.*

How to Copyedit Scientific Books & Journals—*By Maeve O'Connor. ISI Press, 3501 Market Street, Philadelphia, PA 19104. 1986. 150 p. Softcover. ISBN: 0-89495-064-9.*

How to Write & Publish a Scientific Paper, 3rd Edition—*By Robert A. Day. Oryx Press, 4041 North Central at Indian School Road, Phoenix, AZ 85012-3397. 1988. 211 p. Softcover. ISBN: 0-89774-456-X.*

Illustrating Science. *1988. Council of Science Editors, 11250 Roger Bacon Drive Suite 8, Reston VA 20190-5202.*

Language on a Leash—*By Bruce O. Boston. 1988. Editorial Experts, Inc., Alexandria, VA.*

Mark My Words. Instruction and Practice in Proofreading—*By Peggy Smith. Editorial Experts, Inc., 85 S. Bragg Street, Alexandria, VA 22312. 1987. 154 p. (plus exercises and answer keys). Softcover. ISBN: 0-935012-08-7.*

The PC Is Not a Typewriter: A Style Manual for Creating Professional-Level Type on Your Personal Computer—*By Robin Williams Busch. 1991. Peachpit Press. 96p. ISBN: 0-93815-149-5*

Publications Handbook and Style Manual. *1998. ASA–CSSA–SSSA, Madison, WI.*

Scientific English. A Guide for Scientists and Other Professionals—*By Robert A. Day. Oryx Press, 4041 North Central at Indian School Road, Phoenix, AZ 85012-3397. 1992. 125 p. Softcover. ISBN: 0-89774-722-4.*

Scientific Papers and Presentations—*By Martha Davis. 1997. Academic Press, Inc., San Diego, CA.*

Stet! Tricks of the Trade for Writers and Editors—*Edited by Bruce O. Boston. Editorial Experts, Inc., 85 S. Bragg Street, Alexandria, VA 22312. 1986. 310 p. Softcover. ISBN: 0-935012-07-9.*

Writing in Academic Disciplines. Advances in Writing Research, Volume 2—*Edited by David A Joliffe. Ablex Publishing Company, 355 Chestnut Street, Norwood, NJ 07648. 1988. 258 p. Hardcover $67.50. ISBN: 0-89391-434-7.*